PETROLEUM SCIENCE AND TECHNOLOGY

OIL AND CHEMICAL SPILLS

FEDERAL EMERGENCY RESPONSE FRAMEWORK AND RELATED LEGAL AUTHORITIES

PETROLEUM SCIENCE AND TECHNOLOGY

Additional books in this series can be found on Nova's website under the Series tab.

Additional E-books in this series can be found on Nova's website under the E-book tab.

PETROLEUM SCIENCE AND TECHNOLOGY

OIL AND CHEMICAL SPILLS

FEDERAL EMERGENCY RESPONSE FRAMEWORK AND RELATED LEGAL AUTHORITIES

NATALIE CHRISMAN
EDITOR

nova publishers
New York

For permission to use material from this book please contact us:
Telephone 631-231-7269; Fax 631-231-8175
Web Site: http://www.novapublishers.com

Library of Congress Cataloging-in-Publication Data

ISBN: 978-1-62948-840-0

Published by Nova Science Publishers, Inc. † New York

CONTENTS

Preface		**vii**
Chapter 1	Oil and Chemical Spills: Federal Emergency Response Framework *David M. Bearden and Jonathan L. Ramseur*	**1**
Chapter 2	Clean Water Act: A Summary of the Law *Claudia Copeland*	**29**
Chapter 3	Comprehensive Environmental Response, Compensation, and Liability Act: A Summary of Superfund Cleanup Authorities and Related Provisions of the Act *David M. Bearden*	**45**
Chapter 4	Oil Spills in U.S. Coastal Waters: Background and Governance *Jonathan L. Ramseur*	**101**
Chapter 5	Oil Pollution Act of 1990 (OPA): Liability of Responsible Parties *James E. Nichols*	**135**
Index		**161**

PREFACE

This book focuses on the federal emergency response framework of oil and chemical spills; provides a summary of the laws in the Clean Water Act; gives a summary of superfund cleanup authorities and related provisions act; discusses the background and governance of oil spills in United States coastal waters; and finally, provides incite on the liability of responsible parties in the Oil Pollution Act of 1990.

Chapter 1 – Thousands of oil and chemical spills of varying size occur in the United States each year. State and local officials located in proximity to these incidents generally are the first responders and may elevate an incident for federal attention if greater resources are desired. The National Oil and Hazardous Substances Pollution Contingency Plan, often referred to as the National Contingency Plan (NCP), establishes the procedures for the federal response to oil and chemical spills. The scope of the NCP encompasses discharges of oil into or upon U.S. waters and adjoining shorelines and releases of hazardous substances into the environment. Several hundred toxic chemicals and radionuclides are designated as hazardous substances under the NCP, and other pollutants and contaminants also may fall within the scope of its response authorities.

Unlike most federal emergency response plans that are administrative mechanisms, the NCP is codified in federal regulation and is binding and enforceable. The NCP was developed in 1968 and has been revised on multiple occasions to implement the federal statutory response authorities that Congress has expanded over time. Three federal environmental statutes authorized the development of the NCP: the Clean Water Act, as amended; the Comprehensive Environmental Response, Compensation, and Liability Act (CERCLA) of 1980, as amended; and the Oil Pollution Act of 1990.

Several executive orders have delegated the presidential response authorities of these statutes to the federal departments and agencies tasked with implementing the NCP. The lead federal agency serves as the On-Scene Coordinator to direct the resources used in a federal response. The Environmental Protection Agency (EPA) generally is the lead agency responsible for coordinating the federal response within the inland zone, and the U.S. Coast Guard generally serves as the lead agency within the coastal zone. However, a response to an incident occurring on a federal facility is coordinated by the federal department or agency that administers the facility.

The NCP established the National Response System (NRS) as a multi-tiered framework for coordinating the roles of 15 federal departments and agencies that serve as standing members of the National Response Team to offer specialized resources and expertise that the On-Scene Coordinator may call upon to carry out a response. The NRS also outlines the framework for integrating the participation of non-federal entities, including state and local officials, the responsible parties, and other private entities who may wish to contribute resources or expertise.

Although the framework of the NRS is the same for responding to discharges of oil or releases of hazardous substances, the NCP establishes separate operational elements for responding to each type of incident, and these elements differ in some respects. The source of federal funding to carry out a response also differs. The Oil Spill Liability Trust Fund finances the federal response to a discharge of oil, and the Superfund Trust Fund finances the federal response to a release of a hazardous substance. Monies spent from these trust funds may be recouped from the responsible parties under the liability provisions of the Oil Pollution Act and CERCLA, respectively.

For multi-faceted incidents, such as major disasters or emergencies, the NCP also could be invoked under the National Response Framework (NRF) to address an aspect of an incident involving a discharge of oil or release of a hazardous substance. The NRF is a broader administrative mechanism for coordinating the array of federal emergency response plans. However, the NRF itself is not an operational plan that dictates a step-by-step process. The NRF instead merely may apply the NCP as the operational plan to respond to a discharge of oil or release of a hazardous substance.

This report discusses the statutory authorities of the NCP and relevant executive orders; outlines the federal emergency response framework of the NCP to coordinate federal, state, and local roles; and identifies the funding mechanisms to carry out a federal response to a discharge of oil or a release of a hazardous substance.

Chapter 2 – The principal law governing pollution of the nation's surface waters is the Federal Water Pollution Control Act, or Clean Water Act. Originally enacted in 1948, it was totally revised by amendments in 1972 that gave the act its current dimensions. The 1972 legislation spelled out ambitious programs for water quality improvement that have since been expanded and are still being implemented by industries and municipalities.

This report presents a summary of the law, describing the statute without discussing its implementation.

The Clean Water Act consists of two major parts, one being the provisions which authorize federal financial assistance for municipal sewage treatment plant construction. The other is the regulatory requirements that apply to industrial and municipal dischargers. The act has been termed a technology-forcing statute because of the rigorous demands placed on those who are regulated by it to achieve higher and higher levels of pollution abatement under deadlines specified in the law. Early on, emphasis was on controlling discharges of conventional pollutants (e.g., suspended solids or bacteria that are biodegradable and occur naturally in the aquatic environment), while control of toxic pollutant discharges has been a key focus of water quality programs more recently.

Prior to 1987, programs were primarily directed at point source pollution, that is, wastes discharged from discrete sources such as pipes and outfalls. Amendments to the law in that year authorized measures to address nonpoint source pollution (runoff from farm lands, forests, construction sites, and urban areas), which is estimated to represent more than 50% of the nation's remaining water pollution problems. The act also prohibits discharge of oil and hazardous substances into U.S. waters.

Under this act, federal jurisdiction is broad, particularly regarding establishment of national standards or effluent limitations. Certain responsibilities are delegated to the states, and the act embodies a philosophy of federal-state partnership in which the federal government sets the agenda and standards for pollution abatement, while states carry out day-to-day activities of implementation and enforcement.

To achieve its objectives, the act is based on the concept that all discharges into the nation's waters are unlawful, unless specifically authorized by a permit, which is the act's principal enforcement tool. The law has civil, criminal, and administrative enforcement provisions and also permits citizen suit enforcement.

Financial assistance for constructing municipal sewage treatment plants and certain other types of water quality improvements projects is authorized under title VI. It authorizes grants to capitalize State Water Pollution Control Revolving Funds, or loan programs. States contribute matching funds, and under the revolving loan fund concept, monies used for wastewater treatment construction are repaid to states, to be available for future construction in other communities.

Chapter 3 – Congress enacted the Comprehensive Environmental Response, Compensation, and Liability Act of 1980 (CERCLA; P.L. 96-510) in response to a growing desire for the federal government to ensure the cleanup of the nation's most contaminated sites to protect the public from potential harm. The Superfund Amendments and Reauthorization Act of 1986 (P.L. 99-499, SARA) clarified the applicability of the statute's requirements to federal facilities, and modified various response, liability, and enforcement provisions. Several other laws also have amended CERCLA for specific purposes, including relief from cleanup liability for certain categories of parties, and the authorization of federal assistance for the cleanup of abandoned or idled "brownfields" where the presence or perception of contamination may impede economic redevelopment.

CERCLA authorizes cleanup and enforcement actions to respond to actual or threatened releases of hazardous substances into the environment, but generally excludes releases of petroleum and certain other materials covered by other federal laws. Considering the limitation of federal resources to address the many contaminated sites across the United States, CERCLA directs the Environmental Protection Agency (EPA) to maintain a National Priorities List (NPL) to identify the most hazardous sites for the purpose of prioritizing cleanup actions. The states and the public may participate in federal cleanup decisions at NPL sites. The states primarily are responsible for pursuing the cleanup of sites not listed on the NPL, with the federal role at these sites limited mainly to addressing emergency situations.

CERCLA established a broad liability scheme that holds past and current owners and operators of facilities from which a release occurs financially responsible for cleanup costs, natural resource damages, and the costs of federal public health studies. At waste disposal sites, generators of the wastes and transporters of the wastes who selected the site for disposal also are liable under CERCLA. The liability of these "potentially responsible parties" (PRPs) has been interpreted by the courts to be strict, joint and several, and retroactive. At contaminated federal facilities, federal agencies are subject to liability under CERCLA as the owners and operators of those facilities on behalf of the United States. Federal agencies also may be liable in instances in which an agency generated or transported waste for disposal at a non-federal facility.

CERCLA established the Hazardous Substance Superfund Trust Fund to pay for the cleanup of sites where the PRPs cannot be found or cannot pay. A combination of special taxes on industry and general taxpayer revenues originally financed the Superfund Trust Fund, but the authority to collect the industry taxes expired on December 31, 1995. Over time, Congress increased the contribution of general revenues to make up for the shortfall from the expired industry taxes. General revenues now provide most of the funding for the trust fund, but other monies continue to contribute some revenues (i.e., cost-recoveries from PRPs, fines and penalties for violations of cleanup requirements, and interest on the trust fund balance). The availability of these trust fund monies under the Superfund program is subject to appropriations by Congress. Private settlement funds deposited into site-specific Special Accounts within the Superfund Trust Fund also are available to EPA, but are not subject to discretionary appropriations.

Considering the liability of the federal government at its own facilities, the cleanup of federal facilities is not funded with Superfund Trust Fund monies under the Superfund program, but with other federal monies appropriated to the agencies responsible for administering the facilities. However, EPA and the states remain responsible for overseeing and enforcing the implementation of CERCLA at federal facilities to ensure that applicable cleanup requirements are met.

Chapter 4 – The impacts of an oil spill depend on the size of the spill, the rate of the spill, the type of oil spilled, and the location of the spill. Depending on timing and location, even a relatively minor spill can cause significant harm to individual organisms and entire populations. Oil spills can cause impacts over a range of time scales, from days to years, or even decades for certain spills.

Based on data between 1973 and 2009, the annual number and volume of oil spills have shown declines—in some cases, dramatic declines. However, this trend was altered dramatically by the 2010 *Deepwater Horizon* oil spill in the Gulf of Mexico. The incident led to a significant release of oil: according to the federal government's estimate, the well released approximately 206 million gallons of oil before it was contained on July 15. The 2010 Gulf oil spill generated considerable interest in oil spill governance issues.

This report provides background information regarding oil spills in U.S. coastal waters and identifies the legal authorities governing oil spill prevention, response, and cleanup. The governing framework for oil spills in the United States remains a combination of federal, state, and international authorities. Within this framework, several federal agencies have the authority to implement oil spill regulations. Agency responsibilities can be divided into two categories: (1) oil spill response and cleanup and (2) oil spill prevention/preparedness.

Oil spill response authority is determined by the location of the spill: the U.S. Coast Guard has response authority in the U.S. coastal zone, and the Environmental Protection Agency covers the inland zone. Jurisdiction over oil spill prevention and preparedness duties is determined by the potential sources (e.g., vessels, facilities, pipelines) of oil spills.

As with the *Exxon Valdez* oil spill in 1989, the 2010 *Deepwater Horizon* spill generated significant attention to various oil spill policy matters, including prevention, preparedness, response, and liability and compensation. The 111th Congress enacted three oil spill-related proposals into law (P.L. 111-191, P.L. 111-212, and P.L. 111-281), but these laws generally concerned short-term matters that will not have a lasting impact on oil spill governance.

In general, oil spill-related issues garnered less attention during the 112th Congress. The 112th Congress enacted two statutes that contain oil spill-related provisions. P.L. 112-90 includes several oil spill-related provisions involving pipelines. P.L. 112-141 includes a subtitle referred to as the RESTORE Act. This act directs 80% of any administrative and civil Clean Water Act Section 311 penalties connected with the 2010 *Deepwater Horizon* oil spill into a newly created trust fund. Monies from this fund, through various mechanisms, will support environmental and economic restoration in the Gulf states.

Chapter 5 – The Oil Pollution Act of 1990 (OPA) establishes a framework that addresses the liability of responsible parties in connection with the discharge of oil into the navigable waters of the United States, adjoining shorelines, or the exclusive economic zone. Among other provisions, OPA limits certain liabilities of a responsible party in connection with discharges of oil into such areas. The liability limitations established by OPA are currently the subject of significant congressional interest in the wake of the Deepwater Horizon oil spill in the Gulf of Mexico.

A responsible party is strictly liable for removal costs plus damages resulting from an oil spill incident under OPA. A responsible party's liability under OPA is confined to specific categories of damages. Pursuant to OPA, the total liability for damages in connection with an oil spill is limited based on the type of vessel or facility involved, and the amount of oil discharged. OPA provides limited defenses which, if applicable, allow a responsible party to discharge its liability to persons injured by a discharge of oil into the navigable waters of the United States, adjoining shorelines, or the exclusive economic zone.

Claims for removal costs and certain damages must, with limited exception, be presented directly to the responsible party. In the event that a claim for removal costs or certain damages is not paid by the responsible party within 90 days, a claimant may present such a claim directly to the Oil Spill Liability Trust Fund or file suit in court. OPA and its regulations establish procedures for recovering removal costs and damages against the Oil Spill Liability Trust Fund.

This report addresses liability under OPA for removal costs and damages, and the basic procedure for recovering removal costs and damages from the Oil Spill Liability Trust Fund in the event that the responsible party fails to settle such claims.

In: Oil and Chemical Spills
Editor: Natalie Chrisman

ISBN: 978-1-62948-840-0

Chapter 1

OIL AND CHEMICAL SPILLS: FEDERAL EMERGENCY RESPONSE FRAMEWORK*

David M. Bearden and Jonathan L. Ramseur

SUMMARY

Thousands of oil and chemical spills of varying size occur in the United States each year. State and local officials located in proximity to these incidents generally are the first responders and may elevate an incident for federal attention if greater resources are desired. The National Oil and Hazardous Substances Pollution Contingency Plan, often referred to as the National Contingency Plan (NCP), establishes the procedures for the federal response to oil and chemical spills. The scope of the NCP encompasses discharges of oil into or upon U.S. waters and adjoining shorelines and releases of hazardous substances into the environment. Several hundred toxic chemicals and radionuclides are designated as hazardous substances under the NCP, and other pollutants and contaminants also may fall within the scope of its response authorities.

Unlike most federal emergency response plans that are administrative mechanisms, the NCP is codified in federal regulation and is binding and

* This is an edited, reformatted and augmented version of the Congressional Research Service Publication, CRS Report for Congress R43251, dated October 10, 2013.

enforceable. The NCP was developed in 1968 and has been revised on multiple occasions to implement the federal statutory response authorities that Congress has expanded over time. Three federal environmental statutes authorized the development of the NCP: the Clean Water Act, as amended; the Comprehensive Environmental Response, Compensation, and Liability Act (CERCLA) of 1980, as amended; and the Oil Pollution Act of 1990.

Several executive orders have delegated the presidential response authorities of these statutes to the federal departments and agencies tasked with implementing the NCP. The lead federal agency serves as the On-Scene Coordinator to direct the resources used in a federal response. The Environmental Protection Agency (EPA) generally is the lead agency responsible for coordinating the federal response within the inland zone, and the U.S. Coast Guard generally serves as the lead agency within the coastal zone. However, a response to an incident occurring on a federal facility is coordinated by the federal department or agency that administers the facility.

The NCP established the National Response System (NRS) as a multi-tiered framework for coordinating the roles of 15 federal departments and agencies that serve as standing members of the National Response Team to offer specialized resources and expertise that the On-Scene Coordinator may call upon to carry out a response. The NRS also outlines the framework for integrating the participation of non-federal entities, including state and local officials, the responsible parties, and other private entities who may wish to contribute resources or expertise.

Although the framework of the NRS is the same for responding to discharges of oil or releases of hazardous substances, the NCP establishes separate operational elements for responding to each type of incident, and these elements differ in some respects. The source of federal funding to carry out a response also differs. The Oil Spill Liability Trust Fund finances the federal response to a discharge of oil, and the Superfund Trust Fund finances the federal response to a release of a hazardous substance. Monies spent from these trust funds may be recouped from the responsible parties under the liability provisions of the Oil Pollution Act and CERCLA, respectively.

For multi-faceted incidents, such as major disasters or emergencies, the NCP also could be invoked under the National Response Framework (NRF) to address an aspect of an incident involving a discharge of oil or release of a hazardous substance. The NRF is a broader administrative mechanism for coordinating the array of federal emergency response plans. However, the NRF itself is not an operational plan that dictates a step-by-step process. The

NRF instead merely may apply the NCP as the operational plan to respond to a discharge of oil or release of a hazardous substance.

This report discusses the statutory authorities of the NCP and relevant executive orders; outlines the federal emergency response framework of the NCP to coordinate federal, state, and local roles; and identifies the funding mechanisms to carry out a federal response to a discharge of oil or a release of a hazardous substance.

INTRODUCTION

Thousands of oil and chemical spills occur in the United States each year, as a result of accidents, explosions or fires, equipment failure, operator error, and other causes. State and local officials in proximity to an incident usually serve as the first responders to stabilize site conditions and may elevate an incident for federal attention if greater resources are desired. The National Oil and Hazardous Substances Pollution Contingency Plan—often referred to as the National Contingency Plan (NCP) for short—establishes the procedures for the federal response to oil and chemical spills. The scope of the NCP specifically encompasses discharges of oil into or upon U.S. waters and adjoining shorelines and releases of hazardous substances into the environment more broadly. Several hundred toxic chemicals and radionuclides are designated as hazardous substances under the NCP, and other pollutants and contaminants also fall within the scope of its response authorities. The NCP is codified in federal regulation and is authorized in multiple federal statutes. Unlike most other federal emergency response plans that are administrative mechanisms, the regulations of the NCP have the force of law and are binding and enforceable.

After observing the effects of the 1967 *Torrey Canyon* oil tanker spill off the coast of England,[1] the Johnson Administration developed the initial version of the NCP in 1968. The first NCP was an administrative initiative to coordinate the federal response to potential oil spills in U.S. waters. Congress later enacted the Federal Water Pollution Control Act Amendments of 1972[2] (often referred to as the Clean Water Act) to provide explicit statutory authority for the federal response to discharges of oil or hazardous substances into or upon U.S. waters within the contiguous zone and the adjoining shorelines.[3] The 1972 amendments also explicitly directed the preparation of the NCP to carry out these authorities.

The NCP has been revised multiple times since 1968 to implement additional federal statutory authorities that Congress has enacted in the wake of other major incidents. The discovery of severely contaminated sites in the 1970s, such as Love Canal in New York, led to the enactment of the Comprehensive Environmental Response, Compensation, and Liability Act of 1980 (CERCLA).[4] This statute addresses releases of hazardous substances, pollutants, and contaminants into the environment, and it directed a federal response plan for these incidents to be included in the NCP. Although Congress initially considered including oil spills within the response authorities of CERCLA, petroleum was excluded from the scope of the statute with the intention of addressing oil spills in separate legislation.

The 1989 *Exxon Valdez* oil spill in Alaska led to the enactment of the Oil Pollution Act of 1990 (OPA).[5] This statute clarified and expanded the oil spill response authorities of the Clean Water Act extending to U.S. waters within the Exclusive Economic Zone (EEZ)[6] and directed revisions to the NCP to carry out these authorities.

Over time, the NCP has been applied on a routine basis to respond to many varied incidents across the United States involving discharges of oil and releases of hazardous substances. The framework and procedures of the NCP often generate interest among Members of Congress and affected stakeholders in the execution of federal resources to respond to an incident and the participation of state, local, and private entities. Whereas individual incidents may differ in terms of the magnitude, scope, complexity, and associated hazards, effective coordination of the respondents and the adequacy of resources available to carry out a response can be common issues. Larger and more complex spills may garner more prominent attention, such as the 2010 *Deepwater Horizon* oil spill in the Gulf of Mexico. Some raised questions about the authorities and roles of the entities involved in the response to that incident, including federal agencies, state and local governments, and private parties.[7]

This report provides background information on the NCP to address potential questions that may arise in congressional oversight of the federal response to particular incidents. The report discusses the federal statutes that authorize the NCP and related executive orders; mechanisms for reporting incidents to the federal government; the framework under which federal, state, and local roles are to be coordinated; funding mechanisms for federal response actions, including liability for response costs and related damages; and circumstances under which the NCP may be integrated within the National Response Framework (NRF) to address multi-faceted incidents, such as major disasters or emergencies. The **Appendix** to this report provides a chronology of the development of the NCP over time. A list of commonly used acronyms also is provided below.

Table 1. Commonly Used Acronyms and Abbreviations

ACP	Area Contingency Plan
ATSDR	Agency for Toxic Substances and Disease Registry
CDC	Centers for Disease Control and Prevention
CERCLA	Comprehensive Environmental Response, Compensation, and Liability Act
CWA	Clean Water Act
DHS	Department of Homeland Security
EPA	Environmental Protection Agency
HHS	Department of Health and Human Services
NCP	National Contingency Plan (short for National Oil and Hazardous Substances Pollution Contingency Plan)
NPL	National Priorities List
NRF	National Response Framework
NRT	National Response Team
OPA	Oil Pollution Act
OSC	On-Scene Coordinator
RRT	Regional Response Team
SARA	Superfund Amendments and Reauthorization Act of 1986

STATUTORY RESPONSE AUTHORITIES

Since its inception in 1968, the NCP has been revised on multiple occasions to establish regulatory procedures for implementing the federal statutory authorities that Congress has expanded over time to respond to incidents involving a discharge of oil or a release of a hazardous substance. These statutes are outlined briefly below in chronological order of enactment. For a discussion of funding authorized under these statutes to carry out response actions and the scope of liability for response costs and related damages, see the section on "Funding and Liability" later in this report.

The 1972 amendments to the Clean Water Act added Section 311 to the statute to provide explicit statutory authority for the President to respond to a discharge of oil or a hazardous substance into or upon the navigable waters of the United States and adjoining shorelines, and the waters of the contiguous zone.[8] The original NCP had focused on the federal response to oil spills. Section 311 directed the President to further develop the NCP to govern discharges of both oil and hazardous substances within the above aspects of the natural environment. The statute also required several basic elements to be included in the NCP, including mechanisms to coordinate the federal, state, and local roles in responding to an incident, and specific response procedures.

The Comprehensive Environmental Response, Compensation, and Liability Act (CERCLA) of 1980 established the Superfund program and expanded the authorities of the President to respond to releases of hazardous substances into the environment more broadly than Section 311 of the Clean Water Act.[9] Section 101(8) of CERCLA explicitly defines the term "environment" to include "the navigable waters, the waters of the contiguous zone, and the ocean waters of which the natural resources are under the exclusive management authority of the United States under the Magnuson-Stevens Fishery Conservation and Management Act," and any other surface water, groundwater, surface soil, sub-surface soil, or ambient air within or under the jurisdiction of the United States.[10] CERCLA also expanded federal response authority to address releases of pollutants and contaminants into the environment that may present an imminent and substantial danger to the public health or welfare. Section 105 of CERCLA directed the President to revise the NCP to include a plan for responding to these types of incidents.[11] This plan is codified in Subpart E of the regulations of the NCP.[12]

The Superfund Amendments and Reauthorization Act of 1986 (SARA)[13] amended various response, liability, and enforcement provisions of CERCLA. Among these provisions, SARA amended Section 105 of the statute directing the President to further revise the NCP and expand the criteria used to evaluate contaminated sites for placement on the National Priorities List (NPL). The primary purpose of the NPL is to identify—in conjunction with the states—the most threatening sites that warrant federal involvement in long-term remediation. Sites that may require short-term response actions to address emergency situations do not require listing on the NPL to be eligible for federal involvement under the NCP.

The Oil Pollution Act of 1990 expanded and clarified the President's authorities under Section 311 of the Clean Water Act specifically to respond to discharges of oil into or upon the navigable waters of the United States and adjoining shorelines, the waters of the EEZ, or that may affect natural resources belonging to, appertaining to, or under the exclusive management authority of the United States.[14] As amended by OPA, Section 311(d) of the

Clean Water Act directed the President to further revise the NCP to develop specific procedures for implementing these oil spill response authorities.[15] The procedures are codified in Subpart D of the regulations of the NCP and provide the federal plan for undertaking an oil spill response.

EXECUTIVE ORDERS

Several executive orders have delegated the President's response authorities under the above statutes to the federal departments and agencies tasked with implementing the NCP. The two executive orders discussed below amended previous executive orders that initially had delegated the President's response authorities after the enactment of the amendments to the Clean Water Act in 1972 and the enactment of CERCLA in 1980.

- Executive Order 12580 was issued in 1987 and delegated the President's authorities to respond to releases of hazardous substances, pollutants, and contaminants under CERCLA, as amended in 1986 by SARA.[16]
- Executive Order 12777 was issued in 1991 and delegated the President's authorities to respond to discharges of oil under Section 311 of the Clean Water Act, as amended in 1990 by OPA.[17]

Under both executive orders, the coordination of the federal response generally is delegated to EPA within the inland zone and to the U.S. Coast Guard within the coastal zone, unless the two agencies agree otherwise.[18] For the purpose of these delegated roles, the coastal zone is defined in the NCP to include:

> all U.S. waters subject to the tide, United States waters of the Great Lakes, specified ports and harbors on inland rivers, waters of the contiguous zone, other waters of the high seas subject to the NCP, and the land surface or land substrata, ground waters, and ambient air proximal to those waters.[19]

Conversely, the NCP defines the inland zone to encompass "the environment inland of the coastal zone excluding the Great Lakes and specified ports and harbors on inland rivers."[20]

Although responsibility for coordinating a federal response generally is divided between EPA and the U.S. Coast Guard with respect to these two geographic zones, both executive orders assign EPA the sole responsibility of revising the regulations of the NCP, as warranted. Prior to proposing any revisions to the NCP for public comment, EPA must consult with the U.S. Coast Guard and other federal departments and agencies that serve as standing members of the National Response Team (discussed below). As generally is the case with any federal regulation, revisions to the NCP also are subject to the federal rulemaking process, including the opportunity for public comment. Both executive orders further specify that any proposed or finalized revisions to the NCP would be subject to review and approval by the Office of Management and Budget (OMB).

Executive Order 12580 established differing lead agency roles in responding to releases of hazardous substances at federal facilities and vessels. The Department of Defense (DOD) and Department of Energy (DOE) administer most of the federal facilities from which a release of a hazardous substance has occurred.[21] Executive Order 12580 explicitly delegates the President's response authorities under CERCLA to DOD and DOE at facilities and vessels within their respective jurisdiction, custody, or control. DOD also is responsible for responding to incidents involving the removal of military weapons and munitions that are or were under its jurisdiction, custody, or control. Other federal departments and agencies may lead the federal response under CERCLA at facilities and vessels under their jurisdiction, custody, or control only in non- emergency situations. Within their respective zones, EPA or the U.S. Coast Guard would retain the lead role under CERCLA in emergency situations at these federal facilities and vessels.

Executive Order 12777 did not similarly delegate the President's oil spill response authorities under Section 311 of the Clean Water Act with respect to federal facilities and vessels. Under the NCP, the U.S. Coast Guard retains the lead role in responding to discharges of oil from federal facilities or vessels within the coastal zone, regardless of which other federal department or agency may have jurisdiction, custody, or control over that facility or vessel. EPA similarly would lead the response to such incidents at federal facilities in the inland zone. In practice, the federal department or agency that administers the facility or vessel still may carry out a response with its own funds, under the direction of the U.S. Coast Guard or EPA within their respective zones.

The NCP itself also establishes the lead roles of federal departments and agencies in accordance with the delegation of the President's authorities under these two executive orders.[22]

Reporting an Incident

A federal response to a discharge of oil or a release of a hazardous substance may be triggered upon reporting of the incident. The National Response Center serves as the national clearinghouse of all discharges of oil and releases of hazardous substances in the United States that are reported to the federal government.[23] The U.S. Coast Guard is responsible for administering the National Response Center, collecting data on reported incidents, and notifying the appropriate federal departments and agencies that may be involved in responding to an incident under the NCP in coordination with state and local authorities. The U.S. Coast Guard itself generally would be responsible for coordinating a federal response within the coastal zone, but would notify EPA to coordinate the federal response within the inland zone.

The parties who discharge oil or release a hazardous substance are required to report the incident to the National Response Center, if the quantity of the discharge or release exceeds allowable amounts.[24] Reportable discharges of oil include discharges that would (1) violate applicable water quality standards, (2) cause a film or sheen upon or discoloration of the surface of the water or adjoining shorelines, or (3) cause a sludge or emulsion to be deposited beneath the surface of the water or upon adjoining shorelines.[25] Reportable quantities of hazardous substances vary by individual substance in pounds (and kilograms), and in curies (and becquerels) for individual radionuclides designated as hazardous substances.[26] State or local officials, or members of the public, who witness a discharge or release also may report the incident. Furthermore, state or local officials acting as the first responders on-site may contact the National Response Center to elevate a site for federal attention. Once the National Response Center is notified of an incident, a federal response may be undertaken in accordance with the framework and procedures of the NCP to draw upon available resources to respond to potential hazards, discussed next.

Coordination of the Federal Response

The NCP established the National Response System (NRS) as a multi-tiered framework for coordinating the federal response to a discharge of oil or a release of a hazardous substance, pollutant, or contaminant. The NRS establishes the respective roles of federal, state, and local governments in carrying out a federal response, including the party or parties responsible for the incident and other private entities that may wish to contribute resources. As stated in the NCP, the NRS is intended to be "capable of expanding or contracting to accommodate the response effort required by the size or complexity of the discharge or release."[27] Accordingly, the array of respondents and resources used to carry out a response may vary with the magnitude, scope, and complexity of an incident and the associated hazards. The following sections discuss the various components of the NRS, illustrated in **Figure 1** as excerpted from the NCP.

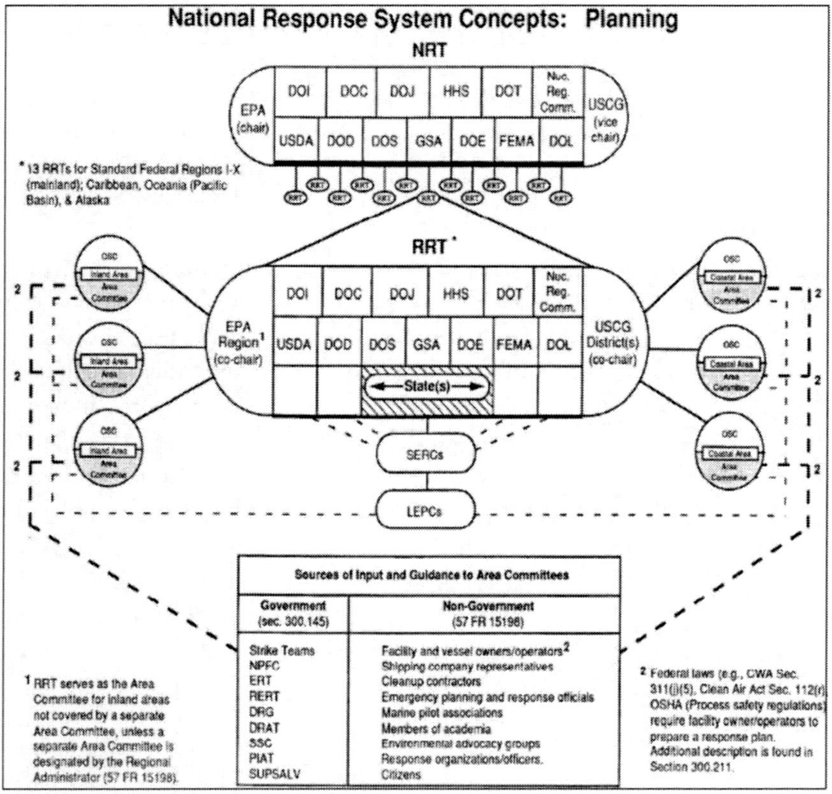

Source: National Oil and Hazardous Substances Pollution Contingency Plan, 40 C.F.R. Part 300, Subpart B— Responsibility and Organization for Response, Section 300.105—General Organization Concepts.

Notes: NRT = National Response Team, RRT = Regional Response Team, OSC = On-Scene Coordinator, SERC = State Emergency Response Commission, LEPC = Local Emergency Response Commission.

Federal Departments and Agencies: EPA = Environmental Protection Agency, USCG = U.S. Coast Guard, DOI = Department of the Interior, DOC = Department of Commerce, DOJ = Department of Justice, HHS = Department of Health and Human Services, DOT = Department of Transportation, Nuc. Reg. Comm. = Nuclear Regulatory Commission, USDA = U.S. Department of Agriculture, DOD = Department of Defense, DOS = Department of State, GSA = General Services Administration, DOE = Department of Energy, FEMA = Federal Emergency Management Agency, and DOL = Department of Labor.

Strike Teams (i.e. specialized teams): NPFC = National Pollution Funds Center, ERT = (EPA) Environmental Response Team, RERT = (EPA) Radiological Emergency Response Team, DRG = (U.S. Coast Guard) District Response Group, DRAT = (U.S. Coast Guard) District Response Advisory Team, SSC = Scientific Support Coordinator, PIAT = Public Information Assist Team, SUPSALV = (U.S. Navy) Supervisor of Salvage.

Figure 1. NCP National Response System.

National Response Team

The National Response Team (NRT) consists of 15 federal departments and agencies (**Figure 1**). The specific role of each department and agency in responding to a discharge of oil or a release of a hazardous substance is outlined in the NCP.[28] These departments and agencies include:

- Environmental Protection Agency (Chair),
- U.S. Coast Guard (Vice-Chair),
- Department of Agriculture,
- Department of Commerce,
- Department of Defense,
- Department of Energy,
- Department of Health and Human Services,
- Department of the Interior,
- Department of Justice,
- Department of Labor
- Department of State,
- Department of Transportation,
- Federal Emergency Management Agency,
- General Services Administration, and
- Nuclear Regulatory Commission.

EPA serves as the "standing" Chair of the NRT, and the U.S. Coast Guard serves as the standing Vice-Chair. Consistent with the delegation of the President's response authorities by executive order, the U.S. Coast Guard becomes the "acting" Chair of the NRT for a federal response to a discharge of oil or a release of a hazardous substance within the coastal zone, and EPA becomes the acting Vice-Chair in such instances.[29] Conversely, EPA remains the Chair for a federal response within the inland zone, and the U.S. Coast Guard the Vice-Chair.

Due to the nature of their ongoing missions, the NRT departments and agencies employ skilled personnel and maintain specialized equipment that can enhance the effectiveness of the federal response. For example, the Department of Health and Human Services (HHS) maintains expertise that may be drawn upon to assess threats to public health resulting from an incident.[30] Within HHS, the Agency for Toxic Substances and Disease Registry (ATSDR) specifically is responsible for assessing public health threats from the release of a hazardous substance, and the Centers for Disease Control and Prevention (CDC) assesses public health threats from discharges of oil. The CDC played a prominent role in assessing threats to public health from the oil spill in the Gulf of Mexico in connection with the *Deepwater Horizon* incident.[31]

Federal departments and agencies that administer federal facilities and vessels are standing members of the NRT to carry out the response to hazardous incidents that may occur in connection with their own activities. Under the NCP, these departments and agencies also may be called upon to use these capabilities in support of the response to non-federal incidents that present similar challenges. For example, the U.S. Navy Supervisor of Salvage maintains specialized equipment and capabilities to respond to pollution incidents involving U.S. Naval vessels. These capabilities may be drawn upon to respond to a civilian incident in ocean environments.[32] During the *Deepwater Horizon* incident, for example, the Navy dispatched oil collection equipment to aid in the federal response under the NCP.

In addition, the Department of Justice also serves as a standing member of the NRT to represent the United States in any litigation that may involve the federal response to a discharge of oil or a release of a hazardous substance under the NCP.[33]

State and Local Involvement in Regional Response Teams

The NCP provides state, territorial,[34] local, and tribal governments the opportunity to participate in the federal response to an incident through the Regional Response Teams that fall under the NRT (**Figure 1**).[35] The NCP established 13 Regional Response Teams.[36] (See **Figure 2** for a map of these regions, as excerpted from the NCP.) Each federal department or agency that is a standing member of the NRT designates an official to serve on each Regional Response Team (RRT) to represent the federal government. The governor of each state or territory within a region may designate an official to represent the state or territorial government. The state and territorial officials serving on a RRT may invite local governments to participate. Indian tribes within a region may designate an official to represent the tribal government.

Because state, territorial, or local officials are likely to be located in closer proximity to incidents that occur within their respective geographic regions, the NCP specifies that they are expected to be the first government representatives on the RRT to arrive at the scene of a discharge or release to take initial response actions.[37] Consequently, state, territorial, or local officials usually are the first responders who may initiate immediate safety measures to protect the public. For example, the NCP indicates that state, territorial, or local officials may be responsible for conducting evacuations of affected populations according to applicable state, territorial, or local procedures. As discussed earlier, state, territorial, or local officials acting as the first responders also may notify the National Response Center to elevate an incident for federal involvement, at which point the coordinating framework of the NCP would be applied.

Enacted in 1986, the Emergency Planning and Community Right-to-Know Act (EPCRA)[38] required each state to create a State Emergency Response Commission (SERC), designate emergency planning districts, and establish Local Emergency Planning Committees (LEPCs) for each district. Each LEPC must prepare a local emergency response plan for the emergency planning district. These local emergency plans are integrated into the appropriate geographic- specific area response plan that may cover several local planning districts, [39] discussed in the "Area Committees" section of this report below.

Area Committees

Area Committees support the Regional Response Teams in preparing for a response to a discharge of oil or a hazardous substance into U.S. waters and the adjoining shorelines, as authorized under Section 311(j)(4) of the Clean Water Act (**Figure 1**).[40] The President may designate "qualified" personnel from federal, state, territorial, and local agencies to serve on these committees. The primary function of each committee is to prepare an Area Contingency Plan (ACP) for its designated geographic area within a region.

Source: National Oil and Hazardous Substances Pollution Contingency Plan (National Contingency Plan or NCP), 40 C.F.R. Part 300, Subpart B—Responsibility and Organization for Response, Section 300.105—General Organization Concepts.

Notes: Alaska (Region X), Hawaii and Pacific Territories (Region IX), and Caribbean Territories (Region II) are treated as separate regions for the purpose of establishing Regional Response Teams under the NCP, resulting in a total of 13 Regional Response Teams, rather than10 teams with respect to the standard 10 federal regions.

Figure 2. Standard State and Territorial Boundaries for Federal Regions.

The geographic-specific aspects of an ACP augment the more general provisions of the NCP. When implemented together, these plans are intended to ensure an effective response to a discharge from a vessel, offshore facility, or onshore facility operating in or near the area. CERCLA does not explicitly authorize Area Committees with respect to a release of a hazardous substance into the environment, whereas Section 311 of the Clean Water Act does authorize such committees to cover discharges of hazardous substances into U.S. waters and the adjoining shorelines. In inland areas not covered by the Clean Water Act, the Regional Response Teams may fulfill the planning functions of the Area Committees.

On-Scene Coordinator

Considering the potentially large number of individuals who may be involved in the federal response to an incident under the NCP, one high-level federal official is responsible for directing and coordinating all of the on-the-ground actions at the scene of a discharge of oil or a release of a hazardous substance. A pre-designated On-Scene Coordinator (OSC) for the geographic area where the discharge or release occurs performs this lead role.[41] Within their respective locales, the OSCs also oversee the development of ACPs by the Area Committees to ensure consistency with the regulatory procedures of the NCP.

EPA generally is responsible for designating the OSCs for incidents involving a discharge of oil or a release of a hazardous substance that may occur in the inland zone, and the U.S. Coast Guard in the coastal zone.[42] U.S. Coast Guard Captains of the Port usually serve as the OSCs, coordinating the response to discharges of oil from all facilities and vessels operating within the coastal zone. The NCP established these lead agency roles in accordance with the executive orders that delegated the President's response authorities, including exceptions for responses to incidents at federal facilities at which the federal department or agency that administers the facility may serve as the OSC instead. See the section of this report on "Executive Orders."

The OSC is responsible for making final decisions on what specific actions are necessary to carry out the federal response to an incident, the use and allocation of federal funds to carry out those actions, what other federal resources may be needed to carry out those actions, and what specific responsibilities are delegated to each entity participating in the federal response, including the party or parties responsible for the incident. The OSC also determines when the federal response to an individual incident is complete and the regulations of the NCP are satisfied. While state, territorial, or local officials may participate in a federal response under the direction of the OSC, the NCP does not preclude states, territories, or local governments from carrying out response actions under their own authorities.

Potential Role of Secretary of Homeland Security

Although EPA and the U.S. Coast Guard usually serve as the OSCs within their respective geographic zones, the Secretary of Homeland Security may assume the lead role in directing a response taken under the NCP in certain circumstances. First, the Secretary of Homeland Security generally has the discretion to assert a lead role in the coastal zone in the capacity of administering the functions of the U.S. Coast Guard within the Department of Homeland Security.[43] Second, Homeland Security Presidential Directive 5 (issued in 2003) more broadly authorizes the Secretary of Homeland Security to be the "principal federal official for domestic incident management" in response to terrorist attacks, major disasters, or other emergencies in any of the following situations:

- a federal department or agency acting under its own authority has requested the assistance of the Secretary;
- the resources of state and local authorities are overwhelmed and federal assistance has been requested by the appropriate state and local authorities;
- more than one federal department or agency has become substantially involved in responding to the incident; or
- the President has directed the Secretary to assume responsibility for managing the domestic incident.[44]

In any of these instances, the procedures and requirements of the NCP still would continue to apply, as the directive is an administrative mechanism that does not preempt existing authorities.[45]

In practice, the Secretary of Homeland Security has not been directly involved on a routine basis in leading response actions taken under the NCP. The lead role of the Secretary generally has been reserved for incidents of greater magnitude, scope, and complexity. For example, Secretary Napolitano coordinated the response taken under the NCP during the 2010 *Deepwater Horizon* oil spill in the Gulf of Mexico. The Secretary's potential role in coordinating a response to other incidents under the NCP generally would depend on the nature of the incident and the need for elevating coordination within the executive branch in the types of situations identified above.

Oil Spills of National Significance

The NCP establishes differing roles with respect to the OSC for an oil spill of national significance (SONS).[46] The NCP does not provide a similar counterpart for releases of hazardous substances.[47] The Administrator of EPA is responsible for designating a SONS in the inland zone, and the Commandant of the U.S. Coast Guard is responsible for making such designations in the coastal zone. For a SONS in the inland zone, the Administrator of EPA may appoint a senior agency official to *assist* the OSC in "communicating with affected parties and the public and coordinating federal, state, local, and international resources at the national level."[48] For a SONS in the coastal zone, the Commandant of the U.S. Coast Guard may appoint a National Incident Commander (NIC) to *assume* the role of the OSC in these capacities, rather than merely assist the OSC. Although the designation of a SONS may affect communication and coordination roles, it does not alter the oil response procedures or requirements of the NCP, and does not make any additional funds available to carry out a response.

In practice, the designation of a discharge of oil as a SONS is rare. On April 29, 2010, Secretary Napolitano classified the *Deepwater Horizon* event as a SONS and appointed U.S. Coast Guard Admiral Thad Allen as the NIC for that incident. This was the first spill to receive such a designation.

Non-Governmental Participation

The participation of non-governmental entities in a federal response may include parties responsible for a discharge of oil or a release of a hazardous substance who perform response actions under the direction of the OSC, private contractors procured either by a responsible party or a federal agency to conduct the physical work, and members of the general public who may wish to contribute resources.[49] In particular, Section 311(j)(5)(D) of the Clean Water Act requires certain types of facilities and vessels to prepare response plans that would ensure the availability of private personnel and equipment to address a worst case discharge of oil or a hazardous substance into or upon navigable waters of the United States, adjoining shorelines, and the Exclusive Economic Zone.[50] These facility and vessel plans must be consistent with the applicable ACPs. In many instances, private facilities and vessels may maintain a contractual relationship with an oil spill removal organization to satisfy this planning requirement for potential oil spills.

The NCP also encourages industry groups, academic organizations, and others to commit resources for federal response operations.[51] Commitments of non-governmental entities may be identified in ACPs, which can be called upon when an incident occurs. Non-governmental entities also may generate scientific or technical information to assist in the development of response strategies, which can be incorporated into ACPs. Individual volunteers also may participate in the federal response. The NCP requires Area Committees to establish procedures that allow for "well organized, worthwhile, and safe use of volunteers."[52] However, the participation of volunteers in the response to a specific incident may be limited to certain activities more appropriate for their skill level or could be restricted if dangerous conditions exist.

FUNDING AND LIABILITY

Congress has established two dedicated trust funds to finance the costs of a federal response to a discharge or oil or release of a hazardous substance, pollutant, or contaminant. Through its National Pollution Funds Center, the U.S. Coast Guard administers the Oil Spill Liability Trust Fund to finance the costs of responding to a discharge of oil.[53] Currently, revenues for the Oil Spill Liability Trust Fund primarily are derived from a dedicated eight cents per-barrel tax on domestic and imported oil. The tax is scheduled to increase to nine cents in 2017 and terminate at the end of that year.

EPA administers the Hazardous Substance Superfund Trust Fund to finance the costs of responding to a release of a hazardous substance, pollutant, or contaminant.[54] The Superfund Trust Fund is financed mostly with revenues transferred from the General Fund of the U.S. Treasury, since the taxes on domestic and imported oil, chemical feedstocks, and corporate income that once financed this trust fund expired at the end of 1995. Neither of these trust funds is available to cover the costs of responding to a discharge of oil or a release of a hazardous substance from a federal facility or vessel. Congress appropriates separate funding directly to the federal departments or agencies with jurisdiction, custody, or control over the facility or vessel from which the discharge or release occurred to pay the response costs of the federal government.

These two trust funds differ in terms of how the monies are made available to carry out a response. Monies from the Oil Spill Liability Trust Fund are authorized as mandatory (i.e., permanent) appropriations that do not require a subsequent discretionary appropriation before they are made available to federal agencies for obligation. However, these monies are subject to certain caps on annual withdrawals from the trust fund[55] and total expenditures per incident.[56] Monies from the Superfund Trust Fund are subject to discretionary appropriations before they are made available for response actions. To enable response capabilities, Congress has annually appropriated monies out of this trust fund to EPA's Superfund account and has reserved separate portions of these funds for emergency "removal" actions versus long-term "remedial" actions. These funds remain available indefinitely until they are expended.

The federal government may recover its response costs from the responsible parties under the liability provisions of OPA and CERCLA, respectively. Recovered funds are to be deposited back into the respective trust fund that financed the federal response.[57] The responsible parties also may perform and pay for response actions up-front with their own monies, subject to direction by the OSC. In the event that the responsible parties are not financially viable or cannot be identified, the applicable trust fund still may pay for federal response actions, up to the amounts made available from that trust fund and within certain limitations.

Financial liability differs somewhat for discharges of oil versus releases of hazardous substances. Section 1002 of OPA establishes the liability of parties responsible for a discharge of oil, including response costs, natural resource damages, certain categories of economic damages, and damages for net costs borne by states and local governments in providing public services in support of a response.[58] Section 107 of CERCLA establishes the liability of parties responsible for a release of a hazardous substance, including response costs, natural resource damages, and the costs of federal public health studies performed by the Agency for Toxic Substances and Disease Registry.[59] Unlike OPA, CERCLA does not establish a separate category of liability for economic damages, with the exception of certain economic losses that may be compensable through natural resource damages based on the loss of the use of resources.[60] Other economic damages attributed to releases of hazardous substances primarily are left to tort law. Notably, CERCLA also does not apply liability to a release of a pollutant or contaminant. The federal government still may respond to such incidents, albeit without a mechanism to recover the costs under CERCLA.

For a presidential declaration of an incident as a major disaster or emergency, funds provided under the Robert T. Stafford Disaster Relief and Emergency Assistance Act (the Stafford Act) may finance the federal response costs,[61] rather than the above trust funds. The use of Stafford Act funds usually would entail the Secretary of Homeland Security applying the NCP under the National Response Framework (discussed below) to address a discharge of oil or a release of a hazardous substance associated with a major disaster or emergency. In practice, the use of Stafford Act funds to pay for the federal response to a discharge of oil or a release of a hazardous substance has been more limited to discharges or releases caused by natural disasters or other emergencies for which there is not a responsible party to pursue. In such instances, the President may make a Stafford Act declaration to provide federal assistance to augment state and local resources, in the absence of viable responsible parties to pay for the response.[62]

NATIONAL RESPONSE FRAMEWORK

The National Response Framework (NRF) is the federal government's broader administrative mechanism that is intended to coordinate the array of federal response plans.[63] As such, the NRF provides the administrative policies and guiding principles for a unified response from all levels of government, and all sectors of communities, to all types of hazards through the combined scope of the various federal response plans that it incorporates. However, the NRF itself is not an operational plan that dictates a step-by-step process for responding to a specific type of hazard, nor is the NRF codified in federal regulation like the NCP.[64]

Through Emergency Support Function (ESF) #10 of the NRF, the Secretary of Homeland Security may apply the operational elements of the NCP for incidents involving the discharge of oil or release of hazardous materials that require a coordinated federal response.[65] (ESF #10 references the term hazardous materials and defines that term to include hazardous substances, pollutants, and contaminants covered under the NCP.) Situations in which the application of the NCP through the NRF may occur include:

- a major disaster or emergency declared under the Stafford Act, when state and local authorities are overwhelmed and federal assistance is requested;
- an incident to which a federal agency is responding under its own authority and requests support from other federal agencies to respond to aspects of the incident that involve the discharge of oil or release of hazardous materials; or

- an incident for which the Department of Homeland Secretary determines that it should lead the response because of special circumstances.

In practice, the federal response to a discharge of oil or a release of a hazardous substance is most often executed under the regulations of the NCP alone, rather than through the coordinating structures of the NRF under ESF #10. The Secretary of Homeland Security's application of the NCP through the NRF appears to be less common and more limited to multi-faceted incidents of greater magnitude, scope, and complexity that may necessitate the coordination of multiple federal response plans. For example, the Department of Homeland Security has stated that the NCP still was applied to the *Deepwater Horizon* oil spill as a stand-alone regulatory authority without involvement of other federal response plans under the NRF.[66] Regardless of whether the NCP is applied as a stand-alone regulatory authority or through the NRF, the procedures for responding to a discharge of oil or release of a hazardous substance are the same because the NCP remains the operative plan in either instance.

APPENDIX. CHRONOLOGY OF THE NCP

Since its inception in 1968, the NCP has been revised on multiple occasions to develop procedures for implementing the federal statutory authorities that Congress has expanded over time to respond to discharges of oil and releases of hazardous substances, pollutants, and contaminants. Major events in the development of the NCP are outlined below.

- **September 1968:** Several departments in the Johnson Administration published the National Multi-Agency Oil and Hazardous Materials Pollution Contingency Plan. This version established a Joint Operations Center, a national reaction team, and regional reaction teams. This first version of the NCP was prepared under an administrative initiative, and some have characterized its legal authority as being "not as straightforward" as subsequent versions codified in federal regulation.[67]
- **April 1970:** The Water Quality Improvement Act of 1970 (P.L. 91-224) directed the President to publish a National Contingency Plan for the removal of oil. The act included specific details such as task forces

at major ports, a national coordination center, and a schedule identifying potential dispersant uses. This act also altered the primary response authority provision, stating that "the President is authorized to act to remove or arrange for the removal of such oil at any time, unless he determines such removal will be done properly by the owner or operator of the vessel, onshore facility, or offshore facility from which the discharge occurs."

- **June 1970:** Pursuant to P.L. 91-224, the Council on Environmental Quality (CEQ) published in the *Federal Register* a National Oil and Hazardous Materials Pollution Contingency Plan.[68] This plan established the National Response Center, National Response Team, Regional Response Team, and the On-Scene Commander roles (later termed On-Scene Coordinator).[69] The plan addressed discharges of oil and "hazardous polluting substances."
- **August 1971:** CEQ made several modifications to the NCP. CEQ changed the name of the plan to the one that still exists today—the National Oil and Hazardous Substances Pollution Contingency Plan. The revised plan established roles for the newly created Environmental Protection Agency (EPA) and National Oceanic and Atmospheric Administration.[70] In particular, the plan designated EPA as the chair of the National Response Team.[71]
- **October 1972:** The Federal Water Pollution Control Act Amendments of 1972 (P.L. 92-500)—often referred to as the Clean Water Act—amended the 1970 statutory authority of the NCP by explicitly requiring it to address hazardous substances as well as oil.[72]
- **August 1973:** Executive Order 11735 delegated presidential authorities pursuant to the 1972 amendments to the Clean Water Act to various federal agencies, including EPA, the Secretary of the Department in which the U.S. Coast Guard is operating, and the Council on Environmental Quality.[73]
- **August 1973:** CEQ published a revised NCP, pursuant to the 1972 amendments to the Clean Water Act and lessons learned from the National Response Team.[74] The NCP was codified in 40 C.F.R. Part 1510.
- **February 1975:** CEQ issued a revised version of the NCP, based on comments regarding the 1973 NCP.[75]
- **December 1977:** The Clean Water Act Amendments of 1977 (P.L. 95-217) required revisions to the NCP to address "imminent threats" of oil and hazardous substance discharges.

- **March 1980:** CEQ revised the NCP based on the 1977 amendments to the Clean Water Act and experiences with several, high-profile spills at that time.[76] The 1980 changes involved, among other things, state participation, local contingency plans, and a national pollution equipment inventory.
- **December 1980:** The Comprehensive Environmental Response, Compensation, and Liability Act of 1980 (CERCLA; P.L. 96-510) required the President to develop, and incorporate into the NCP, procedures for prioritizing and responding to releases of hazardous substances, pollutants, and contaminants into the environment.
- **August 1981:** Executive Order 12316, among other provisions, delegated to EPA the authority to amend the NCP.[77]
- **July 1982:** EPA issued a revised NCP pursuant to CERCLA.[78] The existing NCP structure was largely unchanged. EPA added a new subpart in the regulations with authorities and requirements that specifically addressed hazardous substance response activities.
- **November 1985:** EPA revised the NCP regulations to, among other changes, address "applicable or relevant and appropriate requirements" (ARARs) during response activities.[79]
- **October 1986:** The Superfund Amendments and Reauthorization Act of 1986 (SARA; P.L. 99-499) amended various response, liability, and enforcement provisions of CERCLA and directed the President to revise the NCP to carry out these authorities.[80] Title III of the act (EPCRA) created SERCs and LEPCs.
- **January 1987:** Executive Order 12580 delegated various functions assigned to the President in CERCLA, as amended by SARA.[81]
- **March 1990:** EPA revised the NCP based on Executive Order 12580 and the amendments to CERCLA in SARA, including changes to hazardous substance reporting and response provisions, federal department and agency roles for federal facilities and vessels, and state and public participation.[82]
- **October 1990:** Catalyzed by the 1989 *Exxon Valdez* oil spill in Alaskan waters, Congress enacted the Oil Pollution Act of 1990 (OPA; P.L. 101-380).[83] Among other provisions, OPA provided authority to the President to perform cleanup immediately using federal resources, monitor the response efforts of the spiller, or direct the spiller's cleanup activities.[84] The act also required the President to amend the NCP to establish procedures and standards for responding to worst- case oil spill scenarios.

- **October 1991:** Executive Order 12777 delegated various functions assigned to the President in OPA.[85]
- **September 1994:** EPA revised the NCP based on OPA and its amendments to the Clean Water Act. The modifications to the NCP reflected the revised authorities of the President (delegated to the U.S. Coast Guard or EPA) to direct and/or undertake responses to oil spills. Additions to the NCP also included provisions regarding the National Strike Force Coordination Center, Area Committees, Area Contingency Plans (ACPs), vessel and facility oil spill response planning requirements, and other response elements.

End Notes

[1] The *Torrey Canyon* spill released approximately 35 million gallons of crude oil into the coastal environments of both England and France (National Oceanic and Atmospheric Administration, Oil Spill Case Histories, Summaries of Significant U.S. and International Spills, 1967-2001, 1992). At that time, many considered this spill to be among the worst environmental disasters in history. See Environmental Protection Agency (EPA), National Oil and Hazardous Substances Pollution Contingency Plan Overview, at http://www.epa.gov/osweroe1/content/lawsregs/ncpover.htm.

[2] P.L. 92-500.

[3] The contiguous zone generally is defined as an area that extends from 12 to 24 nautical miles from shore.

[4] P.L. 96-510.

[5] P.L. 101-380.

[6] The Exclusive Economic Zone (EEZ) generally is defined as an area that encompasses the contiguous zone and extends 200 nautical miles seaward of the baseline from which the territorial sea is measured.

[7] See, for example, Thad Allen, *National Incident Commander's Report: MC252 Deepwater Horizon*, October 2010.

[8] 33 U.S.C. § 1321. For a discussion of the Clean Water Act as a whole, see CRS Report RL30030, *Clean Water Act: A Summary of the Law*, by Claudia Copeland.

[9] 42 U.S.C. § 9601 et. seq. For a more in-depth discussion of CERCLA, see CRS Report R41039, *Comprehensive Environmental Response, Compensation, and Liability Act: A Summary of Superfund Cleanup Authorities and Related Provisions of the Act*, by David M. Bearden.

[10] 42 U.S.C. § 9601(8).

[11] 42 U.S.C. § 9605.

[12] Subpart E of the NCP also establishes the procedures for the long-term remediation of a site that may be necessary to protect human health and the environment, once short-term actions are completed to stabilize site conditions.

[13] P.L. 99-499.

[14] 33 U.S.C. § 2701 et. seq. For a more in-depth discussion of the Oil Pollution Act, see CRS Report RL33705, *Oil Spills in U.S. Coastal Waters: Background and Governance*, by Jonathan L. Ramseur.

[15] 33 U.S.C. § 1321(d).

[16] Executive Order 12580, "Superfund Implementation," 52 *Federal Register* 2923, January 29, 1987.

[17] Executive Order 12777, "Implementation of Section 311 of the Federal Water Pollution Control Act of October 18, 1972, as amended, and the Oil Pollution Act of 1990," 56 *Federal Register* 54757, October 22, 1991.

[18] Executive Order 13286 amended both of these Executive Orders to reflect the transfer of the U.S. Coast Guard from the Department of Transportation to the Department of Homeland Security in 2003. See Executive Order 13286, "Amendment of Executive Orders, and Other Actions, in Connection with the Transfer of Certain Functions to the Secretary of Homeland Security," 68 *Federal Register* 10625, 10627, March 5, 2003.

[19] 40 C.F.R. § 300.5.

[20] Ibid.

[21] DOD administers the Defense Environmental Restoration Program to remediate environmental contamination, military munitions, and other safety hazards on active and decommissioned U.S. military facilities. DOE administers the Office of Environmental Management to remediate environmental contamination and dispose of radioactive and other hazardous wastes at former nuclear weapons production sites and federal nuclear research laboratories. The DOE Office of Legacy Management is responsible for the long-term stewardship of federal nuclear facilities that do not have a continuing mission.

[22] 40 C.F.R. § 300.120.

[23] The National Response Center website provides an online reporting tool and contact information to report an incident, presents statistical information on reported discharges and releases over time, and maintains a searchable database of reported incidents, available at http://www.nrc.uscg.mil/nrchp.html.

[24] 40 C.F.R. § 300.300 specifies the requirements and procedures for the discovery or notification of a discharge of oil. 40 C.F.R. § 300.405 specifies the requirements and procedures for the discovery or notification of a release of a hazardous substance. CERCLA generally does not require the reporting of a release of a pollutant or contaminant that is not otherwise designated as a hazardous substance.

[25] 40 C.F.R. Part 110 specifies the criteria for reportable quantities of discharges of oil.

[26] 40 C.F.R. § 302.4 lists the reportable quantity for each hazardous substance designated for the purposes of the NCP.

[27] 40 C.F.R. § 300.5.

[28] 40 U.S.C. § 300.175.

[29] 40 C.F.R. § 300.110.

[30] 40 C.F.R. § 300.175(b)(8).

[31] For information on the CDC's involvement in the federal response to the *Deepwater Horizon* incident, see the agency's website: http://emergency.cdc.gov/gulfoilspill2010.

[32] 40 C.F.R. § 300.175(b)(4)(ii).

[33] 40 C.F.R. § 300.175(b)(10).

[34] References to states in the NCP include U.S. territories and the District of Columbia.

[35] 40 C.F.R. § 300.180.

[36] 40 C.F.R. § 300.115.

[37] 40 C.F.R. § 300.180(f).

[38] EPCRA was in enacted in 1986 as Title III of SARA (42 U.S.C. § 11001-11050).

[39] 40 C.F.R. § 300.210.

[40] 33 U.S.C. § 1321(j)(4).

[41] 40 C.F.R. § 300.120. The NCP uses the term "Remedial Project Manager" instead of "On-Scene Coordinator" for the official designated specifically to coordinate remedial actions. Under CERCLA, "remedial" actions are the long-term actions intended to provide a more permanent means to protect human health and the environment, whereas "removal" actions are intended to address more immediate risks associated with the release of a hazardous substance. When a removal action may precede a remedial action at a site, the On-Scene Coordinator for the initial removal action and the Remedial Project Manager for the

subsequent remedial action are to coordinate their efforts to ensure an "orderly" transition of responsibility.

[42] 40 C.F.R. § 300.120(a).

[43] The U.S. Coast Guard was transferred to the Department of Homeland Security in 2003.

[44] The White House, Management of Domestic Incidents, Homeland Security Presidential Directive 5, Washington, DC, February 28, 2003, available on the Government Printing Office (GPO) website: http://www.gpo.gov/fdsys/pkg/ PPP-2003-book1/pdf/PPP-2003-book1-doc-pg229.pdf.

[45] Section 5 of Homeland Security Presidential Directive 5 states that "Nothing in this directive alters, or impedes the ability to carry out, the authorities of Federal departments and agencies to perform their responsibilities under law."

[46] 40 C.F.R. § 300.323.

[47] As authorized in CERCLA, the NCP establishes procedures for designating sites on the National Priorities List (NPL) for long-term remediation. An NPL designation is not required for conducting emergency responses.

[48] 40 C.F.R. § 300.323(b).

[49] 40 C.F.R. § 300.185.

[50] 33 U.S.C. § 1321(j)(5)(D). The NCP references this provision in 40 C.F.R. § 300.211 and points to specific regulations for different categories of facilities.

[51] 40 C.F.R. § 300.185(a).

[52] 40 C.F.R. § 300.185(c).

[53] 26 U.S.C. § 9509. For additional discussion, see CRS Report R43128, *Oil Sands and the Oil Spill Liability Trust Fund: The Definition of "Oil" and Related Issues for Congress*, by Jonathan L. Ramseur.

[54] 26 U.S.C. § 9507. For additional discussion, see the "Hazardous Substance Superfund Trust Fund" section in CRS Report R41039, *Comprehensive Environmental Response, Compensation, and Liability Act: A Summary of Superfund Cleanup Authorities and Related Provisions of the Act*, by David M. Bearden.

[55] 33 U.S.C. § 2752. The U.S. Coast Guard may "advance" (i.e., withdraw) up to $150 million each fiscal year from the Oil Spill Liability Trust Fund to respond to discharges of oil. Once advanced, the monies remain available until expended. P.L. 111-191 removed this annual limit on the advance of trust fund monies specifically for the *Deepwater Horizon* incident to increase the availability of resources for the federal response. The $150 million annual limit on the advance of trust fund monies remains in place for the federal response to other incidents.

[56] 26 U.S.C. § 9509(c)(2). Total expenditures from the Oil Spill Liability Trust Fund to respond to an individual incident are limited to $1 billion.

[57] The Internal Revenue Code directs recovered costs to be deposited back into the Hazardous Substance Superfund Trust Fund (26 U.S.C. § 9507) and the Oil Spill Liability Trust Fund (26 U.S.C. § 9509), respectively.

[58] 33 U.S.C. § 2702. For additional discussion of liability under OPA, see CRS Report R41266, *Oil Pollution Act of 1990 (OPA): Liability of Responsible Parties*, by Robert Meltz.

[59] 42 U.S.C. § 9607. For additional discussion of liability under CERCLA, see the "Financial Liability" section in CRS Report R41039, *Comprehensive Environmental Response, Compensation, and Liability Act: A Summary of Superfund Cleanup Authorities and Related Provisions of the Act*, by David M. Bearden.

[60] 43 C.F.R. §11.83(c).

[61] 42 U.S.C. § 5121 et seq.

[62] For example, a Stafford Act declaration was not made during the 1989 *Exxon Valdez* oil spill nor the 2010 *Deepwater Horizon* oil spill, as there were viable responsible parties to pursue in each instance.

[63] Authority for the creation of the NRF emanates from numerous sources. FEMA has described the NRF as being guided by 15 "principal emergency authorities," 48 other statutory

authorities and regulations, 17 executive orders, and 20 presidential directives. See U.S. Department of Homeland Security, Federal Emergency Management Agency, National Response Framework: List of Authorities and References, January 2008.

[64] For more information, see CRS Report RL34758, *The National Response Framework: Overview and Possible Issues for Congress*, by Bruce R. Lindsay. For a broad overview of the federal emergency response framework, also see CRS Report R42845, *Federal Emergency Management: A Brief Introduction*, coordinated by Bruce R. Lindsay.

[65] Department of Homeland Security, *National Response Framework—Emergency Support Function #10—Oil and Hazardous Materials Response Annex*, as revised May 2013, available on the Federal Emergency Management Agency website at http://www.fema.gov/media-library-data/20130726-1914-25045-0304/final_esf_10_oil_and_hazardous_materials_response_20130501.pdf.

[66] CRS telephone conversation with Department of Homeland Security Office of Legislative Affairs, July 13, 2010.

[67] Frederick J. Kenney and Melissa A Hamann, "The Flow of Authority to Stop the Flow of Oil: Clean Water Act Section 311(c) Removal Authority and the BP/DEEPWATER HORIZON Oil Spill," *Tulane Maritime Law Journal*, Vol. 36, 2012.

[68] 35 *Federal Register* 8508 (June 2, 1970).

[69] This change was made in the August 1971 rulemaking.

[70] The origin of both the Environmental Protection Agency (EPA) and the National Oceanic and Atmospheric Administration (NOAA) is rooted in a reorganization of the executive branch under the Nixon Administration. The 91st Congress approved this reorganization plan on December 2, 1970. For further details see CRS Report RL30798, *Environmental Laws: Summaries of Major Statutes Administered by the Environmental Protection Agency*, coordinated by David M. Bearden.

[71] A representative from the Department of the Interior served in this role in the previous version of the NCP.

[72] The 1972 act directed EPA to designate hazardous substances in federal regulation.

[73] Executive Order 11735, signed by President Nixon on August 3, 1973, and published in 38 *Federal Register* 21243 (August 7, 1973).

[74] 38 *Federal Register* 21887 (August 13, 1973).

[75] 40 *Federal Register* 6282 (February 10, 1975).

[76] 45 *Federal Register* 17832 (March 19, 1980).

[77] Executive Order 12316, signed by President Reagan August 14, 1981, and published in 46 *Federal Register* 42237 (August 20, 1981).

[78] 47 *Federal Register* 31180 (July 16, 1982).

[79] 50 *Federal Register* 47912 (November 20, 1985).

[80] For further details, see CRS Report R41039, *Comprehensive Environmental Response, Compensation, and Liability Act: A Summary of Superfund Cleanup Authorities and Related Provisions of the Act*, by David M. Bearden.

[81] Executive Order 12580, signed by President Reagan on January 23, 1987, and published in 52 *Federal Register* 2923 (January 29, 1987).

[82] 55 *Federal Register* 8666 (March 8, 1990).

[83] Other oil spills followed the *Exxon Valdez* in 1989 and 1990 (e.g., the *Mega Borg* spilled 5 million gallons of oil in the Gulf of Mexico), further spurring congressional action.

[84] See 33 U.S.C. § 1321(c) as amended by OPA.

[85] Executive Order 12777, signed by President George H.W. Bush on October 18, 1991, and published in 56 *Federal Register* 54757 (October 22, 1991).

In: Oil and Chemical Spills
Editor: Natalie Chrisman

ISBN: 978-1-62948-840-0

Chapter 2

CLEAN WATER ACT: A SUMMARY OF THE LAW*

Claudia Copeland

SUMMARY

The principal law governing pollution of the nation's surface waters is the Federal Water Pollution Control Act, or Clean Water Act. Originally enacted in 1948, it was totally revised by amendments in 1972 that gave the act its current dimensions. The 1972 legislation spelled out ambitious programs for water quality improvement that have since been expanded and are still being implemented by industries and municipalities.

This report presents a summary of the law, describing the statute without discussing its implementation.

The Clean Water Act consists of two major parts, one being the provisions which authorize federal financial assistance for municipal sewage treatment plant construction. The other is the regulatory requirements that apply to industrial and municipal dischargers. The act has been termed a technology-forcing statute because of the rigorous demands placed on those who are regulated by it to achieve higher and higher levels of pollution abatement under deadlines specified in the law. Early on, emphasis was on controlling discharges of conventional

* This is an edited, reformatted and augmented version of a Congressional Research Service publication, CRS Report for Congress RL30030, from www.crs.gov, prepared for Members and Committees of Congress, dated November 30, 2012.

pollutants (e.g., suspended solids or bacteria that are biodegradable and occur naturally in the aquatic environment), while control of toxic pollutant discharges has been a key focus of water quality programs more recently.

Prior to 1987, programs were primarily directed at point source pollution, that is, wastes discharged from discrete sources such as pipes and outfalls. Amendments to the law in that year authorized measures to address nonpoint source pollution (runoff from farm lands, forests, construction sites, and urban areas), which is estimated to represent more than 50% of the nation's remaining water pollution problems. The act also prohibits discharge of oil and hazardous substances into U.S. waters.

Under this act, federal jurisdiction is broad, particularly regarding establishment of national standards or effluent limitations. Certain responsibilities are delegated to the states, and the act embodies a philosophy of federal-state partnership in which the federal government sets the agenda and standards for pollution abatement, while states carry out day-to-day activities of implementation and enforcement.

To achieve its objectives, the act is based on the concept that all discharges into the nation's waters are unlawful, unless specifically authorized by a permit, which is the act's principal enforcement tool. The law has civil, criminal, and administrative enforcement provisions and also permits citizen suit enforcement.

Financial assistance for constructing municipal sewage treatment plants and certain other types of water quality improvements projects is authorized under title VI. It authorizes grants to capitalize State Water Pollution Control Revolving Funds, or loan programs. States contribute matching funds, and under the revolving loan fund concept, monies used for wastewater treatment construction are repaid to states, to be available for future construction in other communities.

INTRODUCTION

The principal law governing pollution of the nation's surface waters is the Federal Water Pollution Control Act, or Clean Water Act. Originally enacted in 1948, it was totally revised by amendments in 1972 that gave the act its current shape. The 1972 legislation spelled out ambitious programs for water quality improvement that have since been expanded and are still being implemented by industries, municipalities and others. Congress made fine-tuning amendments in 1977, revised portions of the law in 1981, and enacted further amendments in 1987.

This report presents a summary of the law, describing the statute. Many details and secondary provisions are omitted here, and even some major components are only briefly mentioned. Further, this report describes the statute, while other CRS products discuss implementation issues.[1] *Table 1* shows the original enactment and subsequent major amendments. *Table 2*, at the end of this report, cites the major U.S. Code sections of the codified statute.

Table 1. Clean Water Act and Major Amendments (codified generally as 33 U.S.C. §§1251-1387)

Year	Act	Public Law
1948	Federal Water Pollution Control Act	P.L. 80-845
		(Act of June 30, 1948)
1956	Water Pollution Control Act of 1956	P.L. 84-660 (Act of July 9, 1956)
1961	Federal Water Pollution Control Act Amendments	P.L. 87-88
1965	Water Quality Act of 1965	P.L. 89-234
1966	Clean Water Restoration Act	P.L. 89-753
1970	Water Quality Improvement Act of 1970	P.L. 91-224, Part I
1972	Federal Water Pollution Control Act Amendments	P.L. 92-500
1977	Clean Water Act of 1977	P.L. 95-217
1981	Municipal Wastewater Treatment Construction Grants Amendments	P.L. 97-117
1987	Water Quality Act of 1987	P.L. 100-4

Authorizations for appropriations to support the law generally expired at the end of FY1990 (September 30, 1990). Programs did not lapse, however, and Congress has continued to appropriate funds to carry out the act. Since the 1987 amendments, although Congress has enacted several bills that reauthorize and modify a number of individual provisions in the law, none comprehensively addressed major programs or requirements.

BACKGROUND

The Federal Water Pollution Control Act of 1948 was the first comprehensive statement of federal interest in clean water programs, and it specifically provided state and local governments with technical assistance funds to address water pollution problems, including research. Water pollution was viewed as primarily a state and local problem, hence, there were no federally required goals, objectives, limits, or even guidelines. When it came to enforcement, federal involvement was strictly limited to matters involving interstate waters and only with the consent of the state in which the pollution originated.

During the latter half of the 1950s and well into the 1960s, water pollution control programs were shaped by four laws that amended the 1948 statute. They dealt largely with federal assistance to municipal dischargers and with federal enforcement programs for all dischargers. During this period, the federal role and federal jurisdiction were gradually extended to include navigable intrastate, as well as interstate, waters. Water quality standards became a feature of the law in 1965, requiring states to set standards for interstate waters that would be used to determine actual pollution levels and control requirements. By the late 1960s, there was a widespread perception that existing enforcement procedures were too time-consuming and that the water quality standards approach was flawed because of difficulties in linking a particular discharger to violations of stream quality standards. Additionally, there was mounting frustration over the slow pace of pollution cleanup efforts and a suspicion that control technologies were being developed but not applied to the problems. These perceptions and frustrations, along with increased public interest in environmental protection, set the stage for the 1972 amendments.

The 1972 statute did not continue the basic components of previous laws as much as it set up new ones. It set optimistic and ambitious goals, required all municipal and industrial wastewater to be treated before being discharged into waterways, increased federal assistance for municipal treatment plant construction, strengthened and streamlined enforcement, and expanded the federal role while retaining the responsibility of states for day-to-day implementation of the law.

The 1972 legislation declared as its objective the restoration and maintenance of the chemical, physical, and biological integrity of the nation's waters. Two goals also were established: zero discharge of pollutants by 1985 and, as an interim goal and where possible, water quality that is both "fishable" and "swimmable" by mid-1983. While those dates have passed, the goals remain, and efforts to attain them continue.

Overview

The Clean Water Act (CWA) today consists of two parts, broadly speaking, one being the Title II and Title VI provisions, which authorize federal financial assistance for municipal sewage treatment plant construction. The other is regulatory requirements, found throughout the act, that apply to industrial and municipal dischargers.

The act has been termed a technology-forcing statute because of the rigorous demands placed on those who are regulated by it to achieve higher and higher levels of pollution abatement. Industries were given until July 1, 1977, to install "best practicable control technology" (BPT) to clean up waste discharges. Municipal wastewater treatment plants were required to meet an equivalent goal, termed "secondary treatment," by that date. (Municipalities unable to achieve secondary treatment by that date were allowed to apply for case-by-case extensions up to July 1, 1988. According to EPA, 86% of all cities met the 1988 deadline; the remainder were put under administrative or court-ordered schedules requiring compliance as soon as possible. However, many cities continue to make investments in building or upgrading facilities needed to achieve secondary treatment, and funding needs remain high; see discussion below.) Cities that discharge wastes into marine waters were eligible for case-by-case waivers of the secondary treatment requirement, where sufficient showing could be made that natural factors provide significant elimination of traditional forms of pollution and that both balanced populations of fish, shellfish, and wildlife and water quality standards would be protected.

The primary focus of BPT was on controlling discharges of conventional pollutants, such as suspended solids, biochemical oxygen demanding materials, fecal coliform and bacteria, and pH. These pollutants are substances that are biodegradable (i.e., bacteria can break them down), occur naturally in the aquatic environment, and deplete the dissolved oxygen concentration in water, which is necessary for fish and other aquatic life.

The act also mandated greater pollutant cleanup than BPT by no later than March 31, 1989, generally requiring that industry use the "best available technology" (BAT) that is economically achievable. BAT level controls generally focus on toxic substances. Compliance extensions of as long as two years are available for industrial sources utilizing innovative or alternative technology. Failure to meet statutory deadlines could lead to enforcement action.

The CWA utilizes both water quality standards and technology-based effluent limitations to protect water quality. Technology-based effluent limitations are specific numerical limitations established by EPA and placed on certain pollutants from certain sources. They are applied to industrial and municipal sources through numerical effluent limitations in discharge permits issued by states or EPA (see discussion of "Permits, Regulations, and Enforcement," below). Water quality standards are standards for the overall quality of water. They consist of the designated beneficial use or uses of a waterbody (recreation, water supply, industrial, or other), plus a numerical or narrative statement identifying maximum concentrations of various pollutants that would not interfere with the designated use. The act requires each state to establish water quality standards for all bodies of water in the state. These standards serve as the backup to federally set technology-based requirements by indicating where additional pollutant controls are needed to achieve the overall goals of the act. In waters where industrial and municipal sources have achieved technology-based effluent limitations, yet water quality standards have not been met, dischargers may be required to meet additional pollution control requirements. For each of these waters, the act requires states to set a total maximum daily load (TMDL) of pollutants at a level that ensures that applicable water quality standards can be attained and maintained. A TMDL is both a planning process for attaining water quality standards and a quantitative assessment of pollution problems, sources, and pollutant reductions needed to restore and protect a river, stream, or lake. Based on state reports, EPA estimates that more than 41,000 U.S. waters are impaired and require preparation of TMDLs.

Control of toxic pollutant discharges has been a key focus of water quality programs. In addition to the BPT and BAT national standards, states are required to implement control strategies for waters expected to remain polluted by toxic chemicals even after industrial dischargers have installed the best available cleanup technologies required under the law. Development of management programs for these post-BAT pollutant problems was a prominent element in the 1987 amendments and is a key continuing aspect of CWA implementation.

Prior to the 1987 amendments, programs in the Clean Water Act were primarily directed at point source pollution, wastes discharged from discrete and identifiable sources, such as pipes and other outfalls. In contrast, except for general planning activities, little attention had been given to nonpoint source pollution (runoff of stormwater or snowmelt from agricultural lands, forests, construction sites, and urban areas), despite estimates that it represents more than 50% of the nation's remaining water pollution problems. As it travels across land surface towards rivers and streams, rainfall and snowmelt runoff picks up pollutants, including sediments, toxic materials, and conventional wastes (e.g., nutrients) that can degrade water quality.

The 1987 amendments authorized measures to address such pollution by directing states to develop and implement nonpoint pollution management programs (Section 319 of the act). States were encouraged to pursue groundwater protection activities as part of their overall nonpoint pollution control efforts. Federal financial assistance was authorized to support demonstration projects and actual control activities. These grants may cover up to 60% of program implementation costs.

The CWA provides for special regulation of the discharge of oil or hazardous substances, because of the potentially catastrophic effects of such events on public health and welfare. Section 311 prohibits the discharge of oil or hazardous substances into U.S. waters. It also requires higher standards of care in the management and movement of oil, including a requirement for spill prevention plans; it enables the government to recover the costs of cleaning up oil and hazardous substance discharges; and it provides for penalties for such discharges. In 1990, Congress enacted the Oil Pollution Act, which partially amended Section 311 and established a comprehensive system for the cleanup of oil spills, adding a mechanism to impose liability for such spills.[2]

While the act imposes great technological demands, it also recognizes the need for comprehensive research on water quality problems. This is provided throughout the statute, on topics including pollution in the Great Lakes and Chesapeake Bay, in-place toxic pollutants in harbors and navigable waterways, and water pollution resulting from mine drainage. The act also authorizes support to train personnel who operate and maintain wastewater treatment facilities.

Federal and State Responsibilities

Under this act, federal jurisdiction is broad, particularly regarding establishment of national standards or effluent limitations. The EPA issues regulations containing the BPT and BAT effluent standards applicable to categories of industrial sources (such as iron and steel manufacturing, organic chemical manufacturing, petroleum refining, and others). Certain responsibilities can be assumed by qualified states, in lieu of EPA, and this act, like other environmental laws, embodies a philosophy of federal-state partnership in which the federal government sets the agenda and standards for pollution abatement, while states carry out day-to-day activities of implementation and enforcement. Responsibilities under the act that may be carried out by qualified states include authority to issue discharge permits to industries and municipalities and to enforce permits; 46 states have been authorized to administer this permit program. EPA issues discharge permits in the remaining states—Idaho, Massachusetts, New Hampshire, New Mexico—the District of Columbia, and most U.S. territories. In addition, as noted above, all states are responsible for establishing water quality standards.

TITLES II AND VI—MUNICIPAL WASTEWATER TREATMENT CONSTRUCTION

Federal law has authorized grants for planning, design, and construction of municipal sewage treatment facilities since 1956 (Act of July 9, 1956, or P.L. 84-660). Congress greatly expanded this grant program in 1972 in order to assist cities in meeting the act's pollution control requirements. Since that time Congress has authorized $65 billion and appropriated more than $88 billion in CWA funds to aid wastewater treatment plant construction. Grants are allocated among the states according to a complex statutory formula that combines two factors: state population and an estimate of municipal sewage treatment funding needs derived from a survey conducted periodically by EPA and the states. The most recent estimate indicated that, as of January 2008, $298 billion more would be required to build and upgrade municipal wastewater treatment plants in the United States and for other types of water quality improvement projects that are eligible for funding under the act, a 17% increase over the previous estimate from four years earlier.

Under the Title II construction grants program established in 1972, federal grants were made for several types of projects based on a priority list established by the states. Grants were generally available for as much as 55% of total project costs. For projects using innovative or alternative technology (such as reuse or recycling of water), as much as 75% federal funding was allowed. Recipients were responsible for non-federal costs but were not required to repay federal grants.

Policymakers have debated the balance between assisting municipal funding needs, which remain large, and the impact of aid programs such as the Clean Water Act's on federal spending and budget deficits. In the 1987 amendments, Congress balanced these competing priorities by extending federal aid for wastewater treatment construction through FY1994, yet providing a transition towards full state and local government responsibility for financing after that date. Grants under the previous Title II program were authorized through FY1990. Under Title VI of the act, grants to capitalize State Water Pollution Control Revolving Funds, or loan programs, were authorized beginning in FY1989 to replace the Title II grants. States contribute matching funds, and under the revolving loan fund concept, monies used for wastewater treatment construction will be repaid to a state fund, to be available for future construction in other communities.

All states now have functioning loan programs, but the shift from federal grants to loans has been easier for some than others. The new financing requirements have been a problem for cities (especially small towns) that have difficulty repaying project loans. Statutory authorization for grants to capitalize state loan programs expired in 1994; however, Congress has continued to provide annual appropriations. An issue affecting some cities is overflow discharges of inadequately treated wastes from municipal sewers and how cities will pay for costly remediation projects. In 2000, Congress amended the act to authorize a two-year $1.5 billion grant program to help cities reduce these wet weather flows. Authorization for that wet weather grant program expired at the end of FY2003 and has not been renewed.

PERMITS, REGULATIONS, AND ENFORCEMENT

To achieve its objectives, the CWA embodies the concept that all discharges into the nation's waters are unlawful, unless specifically authorized by a permit. Thus, more than 65,000 conventional industrial and municipal dischargers must obtain permits from EPA (or qualified states) under the act's National Pollutant Discharge Elimination System (NPDES) program (authorized in Section 402 of the act). NPDES permits also are required for more than 150,000 industrial and municipal sources of stormwater discharges. An NPDES permit requires the discharger (source) to attain technology-based effluent limits (BPT or BAT for industry, secondary treatment for municipalities, or more stringent where needed for water quality protection). Permits specify the effluent limitations a discharger must meet, and the deadline for compliance. Sources also are required to maintain records and to carry out effluent monitoring activities. Permits are issued for no more than five years and must be renewed thereafter to allow continued discharge.

The NPDES permit incorporates numerical effluent limitations issued by EPA. The initial BPT limitations focused on regulating discharges of conventional pollutants, such as bacteria and oxygen-consuming materials. The more stringent BAT limitations emphasize controlling toxic pollutants—

heavy metals, pesticides, and other organic chemicals. In addition to these limitations applicable to categories of industry, EPA has issued water quality criteria for more than 115 pollutants, including 65 named classes or categories of toxic chemicals, or "priority pollutants." These criteria recommend ambient, or overall, concentration levels for the pollutants and provide guidance to states for establishing water quality standards that will achieve the goals of the act.

A separate type of permit is required to dispose of dredged or fill material in the nation's waters, including wetlands. Authorized by Section 404 of the act, this permit program is administered by the U.S. Army Corps of Engineers, subject to and using EPA's environmental guidance. Some types of activities are exempt from permit requirements, including certain farming, ranching, and forestry practices which do not alter the use or character of the land; some construction and maintenance; and activities already regulated by states under other provisions of the act. EPA may delegate certain Section 404 permitting responsibility to qualified states and has done so twice (Michigan and New Jersey). For some time, the act's wetlands permit program has been one of the most controversial parts of the law. Some who wish to develop wetlands maintain that federal regulation intrudes on and impedes private land-use decisions, while environmentalists seek more protection for remaining wetlands and limits on activities that are authorized to take place in wetlands.

Nonpoint sources of pollution, which EPA and states believe are responsible for the majority of water quality impairments in the nation, are not subject to CWA permits or other regulatory requirements under federal law. They are covered by state programs for the management of runoff, under Section 319 of the act.

Other EPA regulations under the CWA include guidelines on using and disposing of sewage sludge and guidelines for discharging pollutants from land-based sources into the ocean. (A related law, the Ocean Dumping Act, 33 U.S.C. §§1401-45, regulates the intentional disposal of wastes into ocean waters.[3]) EPA also provides guidance on technologies that will achieve BPT, BAT, and other effluent limitations.

The NPDES permit, containing effluent limitations on what may be discharged by a source, is the act's principal enforcement tool. EPA may issue a compliance order or bring a civil suit in U.S. district court against persons who violate the terms of a permit. The penalty for such a violation can be as much as $25,000 per day. Stiffer penalties are authorized for criminal violations of the act—for negligent or knowing violations—of as much as $50,000 per day, three years' imprisonment, or both. A fine of as much as

$250,000, 15 years in prison, or both, is authorized for "knowing endangerment"—violations that knowingly place another person in imminent danger of death or serious bodily injury. Finally, EPA is authorized to assess civil penalties administratively for certain well-documented violations of the law. These civil and criminal enforcement provisions are contained in Section 309 of the act. EPA, working with the Army Corps of Engineers, also has responsibility for enforcing against entities who fail to obtain or comply with a Section 404 permit.

While the CWA addresses federal enforcement, the majority of actions taken to enforce the law are undertaken by states, both because states issue the majority of permits to dischargers and because the federal government lacks the resources for day-to-day monitoring and enforcement. Like most other federal environmental laws, CWA enforcement is shared by EPA and states, with states having primary responsibility. However, EPA has oversight of state enforcement and retains the right to bring a direct action where it believes that a state has failed to take timely and appropriate action or where a state or local agency requests EPA involvement. Finally, the federal government acts to enforce against criminal violations of the federal law.

In addition, individuals may bring a citizen suit in U.S. district court against persons who violate a prescribed effluent standard or limitation or permit requirement. Citizens also may bring suit against the Administrator of EPA for failure to carry out a nondiscretionary duty under the act.

SELECTED REFERENCES

Hamilton, Pixie, Timothy Miller, and Donna Myers. "Water Quality in the Nation's Streams and Aquifers—Overview of Selected Findings, 1991-2001." U.S. Geological Survey Circular 1265. May 2004.

Lavelle, Marianne. "Water Woes." U.S. News & World Report. June 4, 2007. pp. 37-53.

U.S. Congressional Budget Office. Public Spending on Transportation and Water Infrastructure. November 2010. Pub. No. 4088. 49 p.

U.S. Environmental Protection Agency. EPA's Draft Report on the Environment, Technical Document. Chapter 2, Purer Water. June 2003. EPA 600-R-03-050. pp. 2-1 - 2-64.

——. Clean Watersheds Needs Survey 2008, Report to Congress. May 2010. EPA-832-R-10-002.

—— Clean Water State Revolving Fund Programs 2009 Annual Report. June 2010. EPA-832-R- 10-001. 36 p.
—— National Water Quality Inventory: Report to Congress for the 2004 Reporting Cycle. January 2009. EPA-841-R-08-001. 37 p.
U.S. Government Accountability Office. Water Quality, A Catalog of Related Federal Programs. GAO/RCED-96-173. June 1996. 64 p.
—— Water Infrastructure: Information on Financing, Capital Planning, and Privatization. GAO-02-764. August 2002. 79 p.

Table 2. Major U.S. Code Sections of the Clean Water Act (codified generally as 33 U.S.C., Chapter 26, Sections 1251-1387)

33 U.S.C.	Section Title	Clean Water Act (as amended)
Subchapter I -	Research and Related Programs	
1251	Declaration of goals and policy	Sec. 101
1252	Comprehensive programs for water pollution control	Sec. 102
1253	Interstate cooperation and uniform laws	Sec. 103
1254	Research, investigations, training and information	Sec. 104
1255	Grants for research and development	Sec. 105
1256	Grants for pollution control programs	Sec. 106
1257	Mine water pollution control demonstrations	Sec. 107
1258	Pollution control in the Great Lakes	Sec. 108
1259	Training grants and contracts	Sec. 109
1260	Applications for training grants or contracts; allocations of grants or contracts	Sec. 110
1261	Award of scholarships	Sec. 111
1262	Definitions and authorizations	Sec. 112
1263	Alaska village demonstration project	Sec. 113
1264	Lake Tahoe study	Sec. 114
1265	In-place toxic pollutants	Sec. 115
1266	Hudson River PCB reclamation demonstration project	Sec. 116
1267	Chesapeake Bay	Sec. 117
1268	Great Lakes	Sec. 118
1269	Long Island Sound	Sec. 119
1270	Lake Champlain Basin program	Sec. 120
1273	Lake Pontchartrain Basin	Sec. 121

33 U.S.C.	Section Title	Clean Water Act (as amended)
1274	Wet weather watershed pilot projects	Sec. 122
Subchapter II -	Grants for Construction of Treatment Works	
1281	Purpose	Sec. 201
1282	Federal share	Sec. 202
1283	Plans, specifications, estimates, and payments	Sec. 203
1284	Limitations and conditions	Sec. 204
1285	Allotment	Sec. 205
1286	Reimbursement and advanced construction	Sec. 206
1287	Authorization	Sec. 207
1288	Areawide waste treatment management	Sec. 208
1289	Basin planning	Sec. 209
1290	Annual survey	Sec. 210
1291	Sewage collection systems	Sec. 211
1292	Definitions	Sec. 212
1293	Loan guarantees for construction of treatment works	Sec. 213
1294	Public information on water recycling, reuse	Sec. 214
1295	Requirements for American materials	Sec. 215
1296	Determination of priority	Sec. 216
1297	Cost-effectiveness guidelines	Sec. 217
1298	Cost effectiveness	Sec. 218
1299	State certification of projects	Sec. 219
1300	Pilot program for alternative water source projects	Sec. 220
1301	Sewer overflow control grants	Sec. 221
Subchapter III -	Standards and Enforcement	
1311	Effluent Limitations	Sec. 301
1312	Water quality related effluent limitations	Sec. 302
1313	Water quality standards and implementation plans	Sec. 303
1314	Information and guidelines	Sec. 304
1315	Water quality inventory	Sec. 305
1316	National standards of performance	Sec. 306
1317	Toxic and pretreatment effluent standards	Sec. 307
1318	Inspections, monitoring, and entry	Sec. 308
1319	Federal enforcement	Sec. 309

Table 2. (Continued)

33 U.S.C.	Section Title	Clean Water Act (as amended)
1320	International pollution abatement	Sec. 310
1321	Oil and hazardous substance liability	Sec. 311
1322	Marine sanitation devices	Sec. 312
1323	Federal facilities pollution control	Sec. 313
1324	Clean lakes	Sec. 314
1325	National study commission	Sec. 315
1326	Thermal discharges	Sec. 316
1327	Financing study	Sec. 317
1328	Aquaculture	Sec. 318
1329	Nonpoint source management programs	Sec. 319
1330	National estuary program	Sec. 320
Subchapter IV -	Permits and Licenses	
1341	Certification	Sec. 401
1342	National pollutant discharge elimination system	Sec. 402
1343	Ocean discharge criteria	Sec. 403
1344	Permits for dredged or fill material	Sec. 404
1345	Disposal of sewage sludge	Sec. 405
1346	Coastal recreation water quality monitoring and notification	Sec. 406
Subchapter V -	General Provisions	
1361	Administration	Sec. 501
1362	General definitions	Sec. 502
1363	Water pollution control advisory board	Sec. 503
1364	Emergency powers	Sec. 504
1365	Citizen suits	Sec. 505
1366	Appearance	Sec. 506
1367	Employee protection	Sec. 507
1368	Federal procurement	Sec. 508
1369	Administrative procedure and judicial review	Sec. 509
1370	State authority	Sec. 510
1371	Other affected authority	Sec. 511
1372	Separability	Sec. 512
1372	Labor standards	Sec. 513
1373	Public health agency coordination	Sec. 514
1374	Effluent standards and water quality information advisory committee	Sec. 515

33 U.S.C.	Section Title	Clean Water Act (as amended)
1375	Reports to Congress	Sec. 516
1376	General authorization	Sec. 517
1377	Indian tribes	Sec. 518
1251 note	Short title	Sec. 519
Subchapter VI -	State Water Pollution Control Revolving Funds	
1381	Grants to states for establishment of revolving funds	Sec. 601
1382	Capitalization grant agreements	Sec. 602
1383	Water pollution control revolving loan funds	Sec. 603
1384	Allotment of funds	Sec. 604
1385	Corrective action	Sec. 605
1386	Audits, reports, and fiscal controls; intended use plan	Sec. 606
1387	Authorization of appropriations	Sec. 607

Note: This table shows only the major code sections. For more detail and to determine when a section was added, the reader should consult the official printed version of the U.S. Code.

End Notes

[1] For example, see CRS Report R41594, Water Quality Issues in the 112th Congress: Oversight and Implementation, by Claudia Copeland.

[2] P.L. 101-380; 33 U.S.C. § 2701 et seq.

[3] CRS Report RS20028, Ocean Dumping Act: A Summary of the Law, by Claudia Copeland.

In: Oil and Chemical Spills
Editor: Natalie Chrisman

ISBN: 978-1-62948-840-0

Chapter 3

COMPREHENSIVE ENVIRONMENTAL RESPONSE, COMPENSATION, AND LIABILITY ACT: A SUMMARY OF SUPERFUND CLEANUP AUTHORITIES AND RELATED PROVISIONS OF THE ACT*

David M. Bearden

SUMMARY

Congress enacted the Comprehensive Environmental Response, Compensation, and Liability Act of 1980 (CERCLA; P.L. 96-510) in response to a growing desire for the federal government to ensure the cleanup of the nation's most contaminated sites to protect the public from potential harm. The Superfund Amendments and Reauthorization Act of 1986 (P.L. 99-499, SARA) clarified the applicability of the statute's requirements to federal facilities, and modified various response, liability, and enforcement provisions. Several other laws also have amended CERCLA for specific purposes, including relief from cleanup liability for certain categories of parties, and the authorization of federal assistance for the cleanup of abandoned or idled

---

* This is an edited, reformatted and augmented version of the Congressional Research Service Publication, CRS Report for Congress R41039, dated June 14, 2012.

"brownfields" where the presence or perception of contamination may impede economic redevelopment.

CERCLA authorizes cleanup and enforcement actions to respond to actual or threatened releases of hazardous substances into the environment, but generally excludes releases of petroleum and certain other materials covered by other federal laws. Considering the limitation of federal resources to address the many contaminated sites across the United States, CERCLA directs the Environmental Protection Agency (EPA) to maintain a National Priorities List (NPL) to identify the most hazardous sites for the purpose of prioritizing cleanup actions. The states and the public may participate in federal cleanup decisions at NPL sites. The states primarily are responsible for pursuing the cleanup of sites not listed on the NPL, with the federal role at these sites limited mainly to addressing emergency situations.

CERCLA established a broad liability scheme that holds past and current owners and operators of facilities from which a release occurs financially responsible for cleanup costs, natural resource damages, and the costs of federal public health studies. At waste disposal sites, generators of the wastes and transporters of the wastes who selected the site for disposal also are liable under CERCLA. The liability of these "potentially responsible parties" (PRPs) has been interpreted by the courts to be strict, joint and several, and retroactive. At contaminated federal facilities, federal agencies are subject to liability under CERCLA as the owners and operators of those facilities on behalf of the United States. Federal agencies also may be liable in instances in which an agency generated or transported waste for disposal at a non-federal facility.

CERCLA established the Hazardous Substance Superfund Trust Fund to pay for the cleanup of sites where the PRPs cannot be found or cannot pay. A combination of special taxes on industry and general taxpayer revenues originally financed the Superfund Trust Fund, but the authority to collect the industry taxes expired on December 31, 1995. Over time, Congress increased the contribution of general revenues to make up for the shortfall from the expired industry taxes. General revenues now provide most of the funding for the trust fund, but other monies continue to contribute some revenues (i.e., cost-recoveries from PRPs, fines and penalties for violations of cleanup requirements, and interest on the trust fund balance). The availability of these trust fund monies under the Superfund program is subject to appropriations by Congress. Private settlement funds deposited into site-specific Special Accounts within the Superfund Trust Fund also are available to EPA, but are not subject to discretionary appropriations.

Considering the liability of the federal government at its own facilities, the cleanup of federal facilities is not funded with Superfund Trust Fund monies under the Superfund program, but with other federal monies appropriated to the agencies responsible for administering the facilities. However, EPA and the states remain responsible for overseeing and enforcing the implementation of CERCLA at federal facilities to ensure that applicable cleanup requirements are met.

INTRODUCTION

By the end of the 1970s, Congress had enacted several environmental laws to regulate sources of pollution in the United States, but had not yet addressed responsibility for contamination resulting from releases of pollutants into the environment. In the late 1970s, the discovery of severely contaminated sites, such as "Love Canal" in New York and Times Beach in Missouri, raised questions as to whether there should be a federal role in cleaning up environmental contamination to protect the public from potential harm. Congress enacted the Comprehensive Environmental Response, Compensation, and Liability Act of 1980 (CERCLA; P.L. 96-510) to authorize the federal government to clean up contaminated sites in the United States and to make the "potentially responsible parties" connected to those sites financially liable for the cleanup costs. CERCLA created the Superfund program to carry out these authorities. The Environmental Protection Agency (EPA) administers the program. Subsequent amendments to CERCLA also authorized EPA to administer a separate grant program to support the cleanup of abandoned or idled "brownfields" properties to encourage their redevelopment.

CERCLA established a broad liability scheme that holds both past and current owners and operators of contaminated facilities financially responsible for the costs of cleanup. At waste disposal sites, generators of the waste sent to the site for disposal, and transporters of the waste who selected the site for disposal, also are responsible for the cleanup costs. If these potentially responsible parties cannot be found or cannot pay for the cleanup, CERCLA authorizes the federal government to finance the cleanup to ensure the protection of human health and the environment. These costs borne by the federal government are referred to as "orphan shares." The broad liability scheme of CERCLA is intended to capture all parties that may have had some involvement in the actions that resulted in contamination of the environment, in order to minimize the burden of the costs of cleanup on the general taxpayer who had no involvement. This approach to liability is based on the principle that polluters should be required to pay for the environmental damage that they cause, often referred to as the "polluter pays principle."

CERCLA established the Hazardous Substance Superfund Trust Fund to finance cleanup actions taken by the federal government at contaminated sites where the potentially responsible parties cannot pay or cannot be found. A combination of special taxes on industry and revenues from the General Fund of the U.S. Treasury initially financed the Superfund Trust Fund, but the authority to collect the industry taxes expired at the end of 1995. As the remaining revenues were expended over time, Congress increased the contribution of general Treasury revenues in an effort to make up for the shortfall from the expired industry taxes. The availability of Superfund Trust Fund monies to finance the cleanup of contaminated sites is subject to appropriations by Congress.

Considering the liability of the federal government as a potentially responsible party at its own facilities, the cleanup of federal facilities is not funded with Superfund Trust Fund monies under the Superfund program, but with other federal monies appropriated for other programs administered by the agencies responsible for these facilities. The Department of Defense (DOD) and the Department of Energy (DOE) administer the cleanup of most contaminated federal facilities. EPA and the states are responsible for overseeing and enforcing the implementation of CERCLA at federal facilities to ensure that applicable requirements are met.

To prioritize cleanup actions, CERCLA directed EPA to establish and maintain a National Priorities List (NPL) of the most contaminated sites in the United States which present the greatest risks to human health and the environment. The NPL includes both non-federal sites and federal facilities that are deemed to present a sufficient level of risk to warrant listing. EPA may require the potentially responsible parties to directly perform or pay for cleanup actions themselves. Alternatively, EPA may clean up a contaminated site up-front with appropriated Superfund monies and later recover those funds from the potentially responsible parties (with the exception of the cleanup of federal facilities which must be funded up-front by the administering agencies). In the event that the potentially responsible parties cannot pay or cannot be found, appropriated Superfund monies may be used to pay the orphan shares of cleanup costs at a site, under a cost-sharing agreement with the state in which the site is located.

The following sections of this report summarize the major cleanup authorities of CERCLA and other relevant provisions of the act. The topics discussed herein include the overall scope and reach of these statutory authorities, the process under which cleanup actions are selected and carried out at individual sites, the financial liability of potentially responsible parties for the costs of cleanup actions, the Superfund Trust Fund that may pay for cleanup actions when the potentially responsible parties cannot pay or cannot be found, enforcement of cleanup liability against the potentially responsible parties to minimize the need for federal tax revenues to finance the cleanup of contaminated sites, the applicability of CERCLA to federal facilities, and federal assistance for the cleanup of brownfields properties.

It should be emphasized that how and to what degree a specific contaminant at an individual site must be cleaned up under CERCLA are not specified in the law itself. The specific actions that are required to clean up contaminants at individual sites are determined on a site-by-site basis. Although CERCLA established a general process for making cleanup decisions, more specific direction is provided in EPA regulation and agency guidance. Other federal agencies that administer the cleanup of federal facilities under CERCLA have developed additional guidance documents that apply to their own respective facilities. Although the statutory authorities upon which federal agencies have based their cleanup regulations and guidance are discussed in this report, the content of these regulations and guidance is not examined here.

As such, this report summarizes selected statutory provisions of CERLCA, but does not discuss agency regulations and guidance that may provide more detailed direction for carrying out cleanup actions at individual sites.

MAJOR AMENDMENTS

Congress has amended CERCLA on numerous occasions to clarify the applicability of the cleanup authorities of the statute, and to provide relief from liability for certain categories of parties who may not have been involved in actions that led to contamination, or who may have contributed only certain quantities or types of waste to a site. Congress also has amended the statute to authorize federal assistance for the cleanup of abandoned or idled "brownfields" properties to encourage their redevelopment. Further, certain amendments have addressed unique cleanup challenges at federal facilities, such as the cleanup of unexploded ordnance on decommissioned military training ranges in the United States, and responsibility for the cleanup of contaminated federal property when it is transferred out of federal ownership.

The Superfund Amendments and Reauthorization Act of 1986 (SARA; P.L. 99-499) clarified that federal facilities are subject to the cleanup requirements of CERCLA to the same extent as non-federal entities, and amended various response, liability, and enforcement provisions of the law. The 1986 amendments also renewed the authorization of appropriations for EPA's Superfund program through FY1991, and established a separate Defense Environmental Restoration Program within the Department of Defense (DOD) to address contamination at active and decommissioned military facilities in the United States.

Title VI of the Omnibus Budget Reconciliation Act of 1990 (P.L. 101-508) extended the authorization of appropriations for EPA's Superfund program through FY1994, and Title XI of that statute extended the authority to collect the special Superfund taxes on industry through December 31, 1995. Although reauthorizing legislation has been introduced in various Congresses, the taxing authority for the Superfund Trust Fund has not been renewed to date, nor has the authorization of appropriations for EPA's Superfund program been extended. Instead, Congress has continued to fund the Superfund program primarily with general Treasury revenues through the annual appropriations process. Congress has annually authorized and appropriated funding for the Defense Environmental Restoration Program each year since its establishment. Most of this funding is supported with general Treasury revenues, with the exception of some revenues generated from the sale or lease of closed military bases which help fund their cleanup.

In 1992, the Community Environmental Response Facilitation Act (P.L. 102-426) amended the federal facility provisions of CERCLA to facilitate the transfer of uncontaminated parcels of surplus federal property on which hazardous substances or petroleum products were not released. Section 334 of the National Defense Authorization Act for FY1997 (P.L. 104-201) further amended CERCLA to allow the transfer of contaminated surplus federal property before cleanup is complete, if assurances are provided to guarantee that the property will be cleaned up to a level that would be suitable for its intended use after transfer.

Other amendments have attempted to address the fairness of the liability scheme of CERCLA, either by limiting or eliminating the liability of certain categories of parties. In 1996, the Asset Conservation, Lender Liability, and Deposit Insurance Protection Act (Subtitle E, Title II, Division A of P.L. 104-

208) amended CERCLA to protect certain fiduciaries and financial lenders from liability. In 1999, the Superfund Recycling Equity Act (Title VI, Appendix I of P.L. 106-113) exempted generators and transporters of recyclable scrap materials from cleanup liability under CERCLA, if the person who received the materials disposed of them instead and the disposal resulted in contamination. There had been some concern that the potential liability of generators and transporters under CERCLA could be a deterrent to recycling.

In 2002, the Small Business Liability Relief and Brownfields Revitalization Act (P.L. 107-118) provided relief from cleanup liability for (1) persons who contributed very small quantities of waste or only municipal solid (i.e. non-hazardous) waste to a site, (2) owners of property that became contaminated merely as a result of migration from a contiguous property owned by another person, and (3) "bona fide" prospective purchasers who otherwise may be hesitant to acquire a contaminated property because of potential cleanup liability once acquiring ownership. The 2002 act also established more specific criteria for exempting "innocent" owners of contaminated property from cleanup liability, if they purchased the property without knowledge of the existing contamination and they had no involvement in actions that led to contamination.

As required by the statute, persons seeking an exemption from liability as a "bona fide" prospective purchaser, contiguous property owner, or "innocent" landowner must have performed "all appropriate inquiry" into the prior uses of the property *before* acquiring ownership, and must take "reasonable steps" *after* acquiring ownership to prevent potentially harmful exposure to environmental contamination. Because of these requirements, such persons still may bear some responsibility for managing contamination on their properties, even though they may be exempt from liability for more extensive cleanup actions that may be taken under CERCLA.

In addition to providing relief from liability for certain categories of parties, P.L. 107-118 authorized federal grants to assist in the cleanup of "brownfields" properties. Brownfields properties typically are abandoned, underutilized, or idled sites where the known or suspected presence of contamination, and the potential for cleanup liability, could be viewed as a deterrent to purchase the property for redevelopment. Brownfields properties tend to be less contaminated than sites listed on the NPL, but may need some cleanup to make them suitable for reuse. EPA originally had established a program in 1993 to provide federal assistance for the cleanup of brownfields properties using the general cleanup authorities of CERCLA as the legal basis for this assistance. P.L. 107-118 provided explicit statutory authority for this purpose, and established a separate Brownfields grant program within EPA, apart from the Superfund program.

Table 1 lists CERCLA as enacted in 1980 and the major amendments to the law noted above.

FEDERAL RESPONSE AUTHORITIES

Section 104(a) of CERCLA specifically authorizes the President to respond to a release (or substantial threat of a release) of a hazardous substance into the environment, or of a pollutant or contaminant which may present an "imminent and substantial danger to the public health or welfare."[1] As authorized by Section 115 of CERCLA,[2] the President delegated the response authorities of CERCLA to EPA and other federal agencies by Executive Order.[3] EPA may respond to releases on the land, and the U.S. Coast Guard may respond to releases into inland river ports and harbors, the Great Lakes, and U.S. coastal waters. If a release were to occur at a federal facility, the agency that administers that facility is authorized to take response actions, subject to oversight and enforcement by EPA and the states in which those facilities are located. Federal funding to carry out response actions under CERCLA is subject to appropriations by Congress.

Notification of a release of a hazardous substance is the action that may trigger a federal response under CERCLA. Section 103(a) requires the party responsible for a release to notify the National Response Center if the quantity of the release exceeds the regulatory limit established for that particular substance.[4] These limits are referred to as "reportable quantities," which are specified in federal regulation.[5] State or local officials, or members of the public, who observe or suspect a release of a hazardous substance also may report the incident. Once a release is reported, the National Response Center is to notify the appropriate federal agency that would be responsible for carrying out the President's response authorities under Section 104(a), and for taking any federal enforcement actions that may be necessary against the parties responsible for the release.

Response actions taken under CERCLA most often entail cleanup activities involving the containment, removal, or treatment of environmental contamination to prevent potentially harmful exposure, but may include the temporary or permanent relocation of potentially exposed individuals if warranted. Congress has excluded certain types of environmental contamination from the response authorities of CERCLA, which may be addressed under other federal environmental laws. These exclusions are provided within the statutory definitions of key terms upon which the response authorities of CERCLA hinge, including the terms "hazardous substance," "pollutant or contaminant," and "release." In addition to these exclusions, Congress has placed general limitations on the extent to which response actions may be taken under CERCLA to address releases of hazardous substances, pollutants, or contaminants in certain situations. In effect, these exclusions and limitations may restrict the applicability or scope of the response authorities of CERCLA at a particular contaminated site.

Table 1. Comprehensive Environmental Response, Compensation, and Liability Act (CERCLA) and Major Amendments to the Statute

Year	Title of Statute	Public Law Number
1980	Comprehensive Environmental Response, Compensation, and Liability Act of 1980	P.L. 96-510
1986	Superfund Amendments and Reauthorization Act of 1986	P.L. 99-499
1990	Omnibus Budget Reconciliation Act of 1990	P.L. 101-508, Title VI, § 6301, Title XI, Subtitle B, Part IV, § 11231
1992	Community Environmental Response Facilitation Act	P.L. 102-426
1996	Asset Conservation, Lender Liability, and Deposit Insurance Protection Act	P.L. 104-208, Division A, Title II, Subtitle E
1996	National Defense Authorization Act for Fiscal Year 1997	P.L. 104-201, § 334
1999	Superfund Recycling Equity Act	P.L. 106-113, Appendix I, Title VI
2002	Small Business Liability Relief and Brownfields Revitalization Act	P.L. 107-118

Petroleum Exclusion

The response authorities of CERCLA do not extend to releases of petroleum. Section 101(14) of CERCLA generally excludes releases of petroleum, including crude oil and any fraction thereof, from the definition of a "hazardous substance" for the purposes of the statute.[6] Section 101(33) does the same for the definition of "pollutant or contaminant."[7] Petroleum releases are covered instead by other statutes. The Oil Pollution Act of 1990 (P.L. 101-380) is the primary federal law that addresses releases of petroleum.[8] Other federal laws also provide authorities to respond to petroleum releases in specific situations.

For example, Section 311(c) of the Clean Water Act authorizes the federal actions to respond to releases of petroleum into or on the navigable waters of the United States and adjoining shorelines.[9] Section 9003(h) of the Solid Waste Disposal Act provides federal response authorities for petroleum leaked from underground tanks.[10] In practice, CERCLA has been applied to the cleanup of some wastes containing petroleum only if the wastes also contained hazardous substances that were not part of the petroleum product itself.[11]

Other Exclusions

Section 101(22) of CERCLA also excludes certain types of releases from the definition of the term "release," thereby removing such releases from the statute's reach.[12] A specific category of nuclear materials is excluded from the definition of release, including "source, byproduct, or special nuclear material" released from a nuclear incident or at certain processing sites. The disposal and cleanup of these materials are subject to the Atomic Energy Act. With the exception of these specific nuclear materials, CERCLA generally applies to the release of radionuclides. In federal regulation, EPA has designated several hundred radionuclides as hazardous substances that are subject to the authorities of CERCLA.[13]

Section 101(22) also excludes three other types of releases from the response authorities of CERCLA: (1) a release that would result in exposure solely within the workplace; (2) emissions from engine exhaust of a motor vehicle, train, aircraft, vessel, or power pumping station; and (3) the "normal" application of fertilizer. There also are certain situations identified in CERCLA in which a party would not be subject to liability, such as the proper application of a registered pesticide product[14] or a federally permitted release of a hazardous substance.[15] However, response authority under the statute generally would remain available to EPA in these two instances, just not the enforcement of liability.

Limitations on Response Actions

Section 104(a)(3) limits the extent to which actions may be taken under CERCLA to respond to releases of hazardous substances, pollutants, or contaminants in certain situations. Response actions generally may not be taken in situations involving (1) releases of naturally occurring substances in their unaltered form; (2) releases from products (such as asbestos) that are part of a residential, business, or community structure or building; or (3) releases into public or private drinking water supplies due to deterioration of supply systems through ordinary use.[16] However, in the event of a public health or environmental emergency declared by the President, CERCLA authorizes response actions to be taken under the statute in any of these three situations, if no other person has the authority and capability to respond in a timely manner.

Prioritization and Procedures for Response Actions

Section 105(a) of CERCLA required the President to develop a National Hazardous Substance Response Plan to establish procedures and standards for prioritizing and responding to releases of hazardous substances, pollutants, and contaminants into the environment. The law directed the President to incorporate these procedures and standards into the National Oil and Hazardous Substances Pollution Contingency Plan (referred to as the National Contingency Plan for short, or NCP). As delegated by the President, EPA promulgated the National Hazardous Substance Response Plan in federal regulation as part of the NCP.[17] These regulations govern any response actions taken under CERCLA.

Consistent with the purpose of the NCP, Section 105(a) of CERCLA also required the President to develop a National Priorities List (NPL) of the most hazardous sites in the United States as an administrative mechanism to prioritize response actions.[18] The President has delegated this task to EPA. The NPL must be updated at least once annually. Section 105(c) primarily requires the use of a Hazard Ranking System (HRS) to determine which sites warrant placement on the NPL.[19] The system scores each site based on certain factors, such as the quantity and nature of hazardous substances; the likelihood of the migration of contamination in groundwater, surface water, and air; and the proximity to human populations and sensitive environments. Because of this range of factors, the severity of contamination alone may not necessarily be sufficient cause to list a site on the NPL. For example, a geographically isolated site with substantial contamination still may not score highly enough on the HRS to warrant placement on the NPL, if the distance from human populations prevents the likelihood of exposure.

In addition to the use of the HRS to evaluate eligibility for listing a site on the NPL, there are two other mechanisms under which EPA also may list a site. First, Section 105(a) allowed each state the one-time opportunity to designate a single site within its borders as the state's highest priority for listing on the NPL. Second, EPA may list a site for which the Agency for Toxic Substances and Disease Registry (ATSDR) has issued a public health advisory, if EPA also determines that the contamination presents a significant public health threat and that its use of "remedial" authority will be more cost-effective than its sole use of "removal" authority without listing the site.[20] As discussed below in the "Scope of Response Actions" section, a site must be listed on the NPL as a condition for the availability of Superfund appropriations to perform remedial actions, but removal actions are not subject to such condition.

EPA has listed over 1,600 sites on the NPL over time, including federal facilities. EPA has deleted over 300 of these sites once EPA determined, in concurrence with the states, that the long-term cleanup objectives had been met. The vast majority of the sites were listed based on EPA's evaluation of the potential risks using the HRS, but some sites have been listed as a result of states designating them as their top priority and as a result of an ATSDR public health advisory.[21]

Scope of Response Actions

CERCLA authorizes two types of response actions: "removal" and "remedial" actions. These terms are defined in Sections 101(23)[22] and 101(24)[23] of CERCLA respectively. Removal does not necessarily mean the physical removal of contamination from the soil, surface water, or groundwater, and remedial actions do not necessarily involve treatment of contamination. Rather, both actions may involve various methods to prevent exposure to contamination, including the relocation of potentially exposed individuals if warranted. It should be noted that the NCP allows remedial actions to be financed with Superfund monies only at sites listed on the NPL, whereas removal actions may be financed with Superfund monies at non-NPL sites to address emergency situations.[24] This restriction is intended to reserve Superfund monies for costlier remedial actions at NPL sites that are thought to present the greatest risks. This funding restriction in the regulations is based on the statutory requirement of Section 105(a) of CERCLA for EPA to prioritize contaminated sites for the purpose of taking remedial actions.

Removal actions tend to be shorter term actions that address more immediate risks, whereas remedial actions tend to be longer term actions that offer a more permanent solution. As such, remedial actions often entail more extensive and costly measures. Because of the typically greater extent and cost of remedial actions, they are subject to more in-depth review in the form of a Remedial Investigation and Feasibility Study (RI/FS). An RI/FS involves an investigation of the contamination to assess potential risks of exposure and a study of the feasibility of remedial alternatives to address those risks. Remedial actions also are subject to public participation requirements under Section 117 of CERCLA.[25] (See the "Public Participation" section of this report.) Removal actions are not subject to a similar degree of review or public comment because of the perceived need for swifter response to address more immediate risks.

Section 104(c)(1) generally restricts the timing of removal actions funded with Superfund monies to one year and the cost to $2 million, with exceptions provided in certain situations.[26] For example, a remedial action may exceed these limitations if the continuance of the removal action would contribute to the remedial action planned at the site. These general timing and cost limitations on removal actions are intended to ensure that removal actions are not pursued on a broader scale as a way to avoid the more in-depth review required of remedial actions.

However, CERCLA does not impose these limitations on a removal action funded by a responsible party with its own funds, nor by a federal agency at a federal facility with dedicated monies appropriated to that agency for that purpose apart from Superfund. From a practical standpoint, imposing the above timing and cost limitations on removal actions at many federal facilities administered by the Department of Defense and Department of Energy could constrain the needed scope of removal actions, as cleanup challenges are often greater at these federal facilities in comparison to non-federal sites.

Federal-State Cost Sharing

Section 104(c)(3) of CERCLA requires the state in which a non-federal NPL site is located to agree to share the costs of remedial actions at that site, as a condition of obligating federal Superfund monies to finance those actions.[27] States are not responsible for sharing the costs of cleanup at sites where the potentially responsible parties pay for the cleanup, including federal facilities that are funded by the federal agencies that administer them. Rather, the federal government and the states are to share the costs of assuming the responsibility for the orphan shares of the cleanup costs, for which there are no viable parties to pursue.

This cost-sharing requirement in Section 104(c)(3) is intended to reduce the financial burden on the federal taxpayer presented by the often long-term financial commitment involved in carrying out a remedial action. Notably, CERCLA does not require states to agree to share the costs of removal actions, which typically are less costly as a result of their smaller scope. Consequently, federal Superfund monies may be used to finance the entire costs of removal actions.

At a site where the state must agree to share the costs of remedial actions as a condition of the obligation of federal Superfund monies, the state first must provide certain assurances of its financial commitments, specified in a binding contract or cooperative agreement with the federal government. Absent such contract or agreement, federal Superfund monies are not available to finance remedial actions at that site. To allow the obligation of federal Superfund monies to commence the remedial actions, the state must agree to pay 10% of the costs of those actions. If the site was owned or operated by the state, or a political subdivision of the state, at the time of disposal, the state must agree to pay at least 50% of the costs of the remedial actions.

In addition to the above conditions, the state must agree to perform future maintenance of the remedial actions for their expected operational life. The point of maintenance usually occurs after any necessary construction is complete and the remedial action is operating as intended. CERCLA authorizes a delay in the state's responsibility for the maintenance of groundwater or surface water remedies. Section 104(c)(6) allows a state to delay its maintenance responsibilities for the first 10 years of the operation of such remedial actions.[28] The statute allows a delay in the state's maintenance responsibility specifically for these types of actions to reduce the burden of those costs on the state, as the cleanup of groundwater or surface water tends to be more costly than other types of remediation. During the initial 10-year period, federal Superfund monies instead can be used to pay the maintenance costs of groundwater or surface water remedies.

SELECTION OF RESPONSE ACTIONS

Section 121(a) of CERCLA generally requires response actions at contaminated sites to achieve acceptable levels of exposure that would be protective of human health and the environment.[29] Response actions also are to be cost-effective over both the short-term and long-term, including the operation and maintenance of the action. Section 121(b) states a preference for the selection of remedial actions that involve treatment to "permanently and significantly" reduce the "volume, toxicity or mobility" of contamination, as opposed to actions that do not involve such treatment.[30]

Actions not involving treatment often entail the containment of wastes on-site, or the removal and disposal of wastes off-site. The containment of wastes on-site could present lingering health and environmental risks if the containment method were to fail over time. If the remedial action would result in wastes being left on-site, Section 121(c) requires the President to review the performance of the remedial action every five years to determine whether that action continues to be protective of human health and the environment.[31] If the action is not functioning as intended, the President may take additional remedial actions at the site to achieve the cleanup goal.

Although Section 121 includes certain requirements to govern the selection of remedial actions, it does not specify how clean an individual site must be to protect human health and the environment. Section 121 also does not identify the specific nature of the remedial actions that would be required to attain a cleanup goal at an individual site. Instead, these cleanup decisions are made on a site-by-site basis taking many factors into consideration, including the potential for human exposure based on the anticipated land use, and the technical and economic feasibility of cleanup alternatives to prevent exposure.

Cleanup Standards

The level of cleanup that is required can vary widely from site to site depending on the contaminants present, the cleanup standards or criteria that apply to those contaminants, and the response actions selected to attain those standards or criteria. Rather than specify standards or criteria for individual hazardous substances, Section 121(d) of CERCLA broadly requires that cleanup comply with applicable, relevant, and appropriate requirements (ARARs) to protect human health and the environment.[32] ARARs can include a host of federal or state standards, requirements, or other criteria. In this sense, CERCLA functions as an "umbrella" statute under which other statutes or regulations also may be applied to the cleanup of a contaminated site.

Section 121(d)(4) authorizes the waiver of a particular standard, if:

- the contemplated response action would be part of a larger remedial action that would meet the standard once the larger action is completed;
- compliance with the standard would result in a greater risk than the alternatives;
- compliance with the standard would be technically impracticable from an engineering perspective;
- an equivalent standard of performance would be attained;
- in the case of a state standard, the state has not consistently applied that standard elsewhere within its jurisdiction; or
- meeting the standard would not provide a balance between the need for protection of public health and welfare and the environment at the site under consideration, and the availability of monies in the Superfund Trust Fund to respond to more immediate risks at other sites.[33]

Although CERCLA generally does not list specific standards that may apply to the cleanup of an individual site, there are two sets of standards cited in Section 121(d) that broadly apply to the selection of remedial actions at any site. First, the law requires remedial actions to achieve a level of cleanup that would attain Maximum Contaminant Levels (MCLs) established for current or potential sources of drinking water under the Safe Drinking Water Act.[34] Second, remedial actions must be consistent with other water quality criteria established under Sections 303[35] or 304[36] of the Clean Water Act. However, the applicability of these sets of standards to an individual site remains limited to circumstances in which the standards still are deemed "relevant and appropriate," consistent with the underlying premise of an ARAR.

State Participation

CERCLA authorizes a broad role for states to participate in the cleanup process. States must agree to share in the costs of remedial actions at non-federal NPL sites as a condition of the obligation of federal Superfund monies.

In acknowledgment of their sharing of the costs of cleanup, Section 121(f) of CERCLA requires that states be afforded opportunities for "substantial and meaningful involvement" in initiating, developing, and selecting remedial actions.[37]

However, there are certain limitations on the involvement of states in cleanup decisions at federal facilities, as states do not share in the costs of cleanup at these facilities. If a state wishes to challenge a remedial decision of a federal agency at a facility which that agency administers, Section 121(f)(3) requires that the state show that the decision of the agency is not supported by "substantial evidence."[38]

Public Participation

CERCLA also provides a role for the general public in commenting on the selection of remedial actions at individual sites. This role is similar to that under many other federal laws that require the opportunity for the public to comment on certain types of federal decisions. Section 117 of CERCLA requires EPA, or other federal agency responsible for administering and funding the cleanup of a contaminated site, to provide the public an opportunity to comment on proposals for the selection of remedial actions.[39] Once a final decision is made, public notice of the decision must be provided, with an explanation of any "significant" differences from the proposed action and a response to each "significant" public comment on the proposed action.

The opportunity for public comment required by Section 117 of CERCLA applies only to decisions on remedial actions. Decisions on removal actions are not subject to these requirements because of the presumed need for expedited action to address more immediate risks. In practice, EPA and other federal agencies typically notify the public of the selection of removal actions to inform communities of the nature and timing of such actions. To assist the public in understanding technical information presented in cleanup decision documents, Section 117(e) of CERCLA authorizes technical assistance grants of up to $50,000 for community groups.[40] These grants are available only to affected communities at sites listed on the NPL.

Agency for Toxic Substances and Disease Registry

Section 104(i) of CERCLA established the Agency for Toxic Substances and Disease Registry (ATSDR) primarily to assess potential health risks at NPL sites.[41] The ATSDR assesses individual sites based on the likelihood of human exposure to contamination through the air, soil, surface water, groundwater, and other pathways such as consumption of contaminated food sources. The purpose of these assessments is two-fold: to inform the public of potential health hazards at a contaminated site, and to aid decision-makers in evaluating what cleanup actions may be warranted to prevent potentially harmful exposure. Although the findings of the ATSDR may be used to inform the selection of cleanup actions, the agency does not have any authority to dictate cleanup decisions. In addition to site-specific assessments, Section 104(i) directs the ATSDR to prepare toxicological profiles of hazardous substances commonly found at NPL sites to identify potential health effects that can result from exposure.

Section 104(i) of CERCLA also authorizes the ATSDR to carry out several other functions intended to protect public health. For example, the agency is authorized to provide medical care and testing to individuals in the event of a public health emergency caused by, or believed to be caused by, exposure to toxic substances. CERCLA does not provide any criteria as to what constitutes a public health emergency for this purpose, presumably leaving the declaration of such an emergency to the discretion of the ATSDR. As with other roles, the resources of the agency to fulfill this role are subject to appropriations by Congress. To date, the ATSDR has not used its authority under CERCLA to declare a public health emergency. In practice, the agency's role has focused on educating the public about known health risks from exposure to hazardous substances, and assessing potential risks at individual sites to aid in informing cleanup decisions.

Financial Liability

Section 107 of CERCLA identifies the categories of potentially responsible parties connected with a contaminated site who are liable for the costs of response actions that EPA deems necessary to protect human health and the environment.[42] Such parties also are liable for damages for injury to, destruction of, or loss of natural resources resulting from a release of a hazardous substance, including the costs of assessing such injury, destruction, or loss; and the costs of public health assessments carried out by the ATSDR under Section 104(i) of CERCLA. The following sections discuss the categories of parties who are liable under Section 107 of CERCLA, the reach of liability, defenses to liability, and limitations on the liability of certain categories of parties.

Categories of Potentially Responsible Parties

Section 107(a) identifies four categories of potentially responsible parties who are liable for the costs of response actions, natural resource damages, and public health assessments associated with the release or threatened release of a hazardous substance:

- any person who currently owns or operates a facility or vessel from which a hazardous substance was released;
- any person who at the time of disposal of a hazardous substance owned or operated the facility at which such disposal occurred;
- any person who arranged for the disposal or treatment of a hazardous substance (often referred to as a generator of waste), and any person who arranged for the transport of a hazardous substance for disposal or treatment; and
- any person who accepts or accepted a hazardous substance for transport to a disposal or treatment facility, incineration vessel, or site selected by such person.[43]

In the context of liability, it should be noted that financial responsibility for cleanup costs may extend to actions beyond a facility boundary, if a hazardous substance were to migrate (i.e., move or spread) through the environment. Section 101(8) of CERCLA defines the term "environment" to include not only the land, but also surface water, groundwater, or ambient air.[44] Consequently, cleanup actions may be necessary not only on the facility where the initial release occurred, but anywhere the hazardous substance may migrate through the environment. For example, hazardous substances that migrate into groundwater or surface water can travel some distance, even miles, and can necessitate cleanup actions across a larger area than where the release first occurred.

Reach of Liability

Over time, the courts have interpreted liability under Section 107 of CERCLA to be strict, joint and several, and retroactive. This judicial interpretation is rooted in case law, legislative history, and the definition of liability in Section 101(32) of CERCLA[45] that applies the same standards of liability as in Section 311 of the Clean Water Act.[46]

- Strict liability means that a party can be held liable regardless of whether the conduct of that party was negligent.
- Joint and several liability means that one or more of the liable parties can be held responsible for the full cost of the cleanup at a site, regardless of the degree of involvement in the contamination. However, Section 113(f)(1) of CERCLA allows a party to seek recovery of some of its cleanup costs from other parties at a site through contribution claims in court.[47] In deciding such claims, a court is to base the allocation of cleanup costs on "equitable factors." In the event that a party can show that the waste *it* sent to the site could not have contributed to the contamination, joint and several liability is not to apply to that party.
- Retroactive liability means that parties are liable for the cleanup of hazardous substances released prior to the enactment of CERCLA on December 11, 1980. However, Section 107(f)(1) extends liability for natural resource damages only to releases that occurred on or after the enactment of CERCLA, which resulted in injury to, destruction of, or loss of the natural resources.[48]

It should be emphasized that the above description of the basic liability standards of CERCLA merely offers a brief summary of the broad reach of the statute, as generally interpreted by the courts over time. As such, this description does not examine the complexities of individual court decisions on these matters. Since the enactment of CERCLA in 1980, well over one thousand court decisions have interpreted these basic liability standards under the statute to determine the financial responsibility of potentially responsible parties for the costs of cleanup. How a court may view the cleanup liability of an individual party at any one site would depend on numerous legal issues that are beyond the scope of the summary of CERCLA offered in this report.

Defenses to Liability

Section 107(b) of CERCLA provides defenses to liability under certain circumstances.[49] A party cannot be held liable for the release or threatened release of a hazardous substance, and resulting injury to, destruction of, or loss of natural resources, if that party can provide evidence that the release or threatened release was caused solely by

- an act of God;
- an act of war;
- an act or omission of a third party with whom the defendant has no contractual relationship, if the defendant exercised due care with respect to the hazardous substance and took precautions against foreseeable acts or omissions of that third party and against the foreseeable consequences of such acts or omissions; or
- any combination of these three circumstances.

The third party defense sometimes is characterized as the "innocent" landowner defense, in the sense that it typically pertains to property owners who had no involvement in the actions that led to the contamination. Section 101(35) of CERCLA defines the term contractual relationship for the purpose of the third party defense, and specifies the conditions that a landowner must satisfy to claim the lack of a contractual relationship connecting the owner to the contamination.[50] See the "Bona Fide Prospective Purchasers and Innocent Landowners" section of this report below.

Limitations on Liability

To address the fairness of the liability scheme of CERCLA, Congress has amended the statute at various times to limit, or in some cases eliminate, the liability of certain categories of parties who may not have been involved in actions that resulted in contamination, who may have contributed only very small quantities or less toxic wastes to a contaminated site, or whose conduct Congress did not wish to discourage. These categories of parties include:

- response action contractors who merely perform the work to clean up a contaminated site, but who did not cause or otherwise contribute to the contamination;

- state and local governments that acquired contaminated property involuntarily through bankruptcy, tax delinquency, abandonment, or other circumstances, and did not cause or otherwise contribute to the contamination;
- persons who only hold a contaminated property in a fiduciary capacity;
- financial lenders who acquire financial interests or ownership of a contaminated property through foreclosure;
- generators and transporters of scrap materials intended for recycling, but instead may have been disposed of by other persons;
- persons who contributed only very small quantities of waste or only municipal solid (i.e. non-hazardous) waste to a site;
- service station dealers who only disposed of recycled oil that was not contaminated with hazardous substances, and who fully complied with federal regulations for managing the recycled oil;
- "innocent" landowners who purchased a property without knowledge of existing contamination, with respect to the third party defense noted above;
- other "innocent" owners of property that became contaminated merely through migration from a contiguous property where the initial release occurred; and
- "bona fide" prospective purchasers who otherwise may be hesitant to acquire a property on which contamination is known or suspected to be present, because of the potential liability for cleanup upon acquiring ownership.

Amendments to CERCLA that provided such limitations on cleanup liability for specific categories of parties are examined further below. As discussed in the "Other Exclusions" section of this report, there also are certain situations identified in CERCLA in which a party would not be subject to liability, such as the proper application of a registered pesticide product[51] or a federally permitted release of a hazardous substance.[52]

Cleanup Contractors

Soon after the enactment of CERCLA in 1980, it was realized that a private contractor hired to clean up a contaminated site may be exposed to potential liability as an operator of that site, or as a person who arranged for disposal or transport of waste in instances in which the contractor removed waste as part of the cleanup. This exposure to potential liability was viewed as a deterrent to private contractors being willing to clean up contaminated sites. In response to this concern, Section 119 of the Superfund Amendments and Reauthorization Act of 1986 added Section 119 to CERCLA to limit the liability of "response action" contractors who are hired to perform cleanup actions.[53]

Section 119(a)(1) states that cleanup contractors shall not be liable under CERCLA, or any other federal law, to any person for "injuries, costs, damages, expenses, or other liability" resulting from the release or threatened release of a hazardous substance, pollutant, or contaminant.[54] (However, no immunity from liability under state law is conferred under CERCLA.) Section 119(a)(2) states that a cleanup contractor shall not be exempt from federal liability for a release caused by that contractor as a result of conduct that is "negligent, grossly negligent, or which constitutes intentional misconduct."[55] Under certain circumstances, Section 119(c) authorizes the President the discretion to indemnify a cleanup contractor for negligent conduct, but not grossly negligent conduct or intentional misconduct.[56] Such indemnification is intended to cover a contractor's liability that cannot be covered by insurance at a "fair and reasonable" price.

Fiduciaries and Financial Lenders

Enacted in the 104th Congress, the Asset Conservation, Lender Liability, and Deposit Insurance Protection Act of 1996 (P.L. 104-208, Division A, Title II, Subtitle E of the Omnibus Consolidated Appropriations Act for FY1997) added Section 107(n) to CERCLA to limit the liability of persons who hold a facility or vessel only in a fiduciary capacity for another person to the value of the assets held on behalf of that person.[57] This limitation on liability is provided if the fiduciary did not cause or contribute to a release or threatened release of a hazardous substance from such facility or vessel. Fiduciaries who held an interest in a contaminated property had been concerned that joint and several liability under CERCLA could result in their financial liability exceeding the value of the assets held.

P.L. 104-208 also amended the definition of "owner or operator" in Section 101(20) of CERCLA[58] to exclude financial lenders that did not participate in the management of a facility or vessel from which there was a release or threatened release of a hazardous substance, but who held indicia of ownership primarily to protect security interests. Lenders also were exempted from liability as owners or operators of foreclosed properties, but only if they did not participate in the management of the facility or vessel prior to foreclosure. Lenders especially had been concerned about becoming liable for the cleanup of contaminated properties following foreclosure, when they become owners of the property, and hence could become liable for cleanup under Section 107.

Generators and Transporters of Recyclable Materials

Enacted in the 106th Congress, Title VI—Superfund Recycling Equity—of Appendix I of the Consolidated Appropriations Act for FY2000 (P.L. 106-113) added Section 127 to CERCLA to exempt certain parties involved in the recycling of scrap materials from cleanup liability as generators and transporters of wastes.[59] The exemption is available to persons who "arranged" for the recycling of scrap materials (by selling the materials or otherwise arranging for their recycling). Recyclers involved in these activities had been concerned about becoming liable as generators or transporters of wastes if they sold or transported scrap materials to a facility that disposed of the materials instead of recycling them as intended. Some had perceived this potential liability as a deterrent to recycling.

The exemption is available to the above persons only for materials that fall within the statutory definition of recyclable materials in CERCLA. Section 127(b) defines recyclable materials to include the following scrap materials: plastic, glass, textiles, rubber (other than whole tires), metal, or spent lead-acid, spent nickel-cadmium, and other spent batteries, and minor amounts of material incident to or adhering to the scrap material as a result of its normal and customary use prior to becoming scrap.[60] Two items are expressly excluded: (1) shipping containers of a certain capacity that contained a hazardous substance or onto which a hazardous substance adhered, and (2) materials containing polychlorinated biphenyls (PCBs) in excess of federal standards.

Even if a material can be considered recyclable within the above statutory definition, the exemption is not automatic. Section 127(c) requires the person who arranged for the recycling of the materials to demonstrate that certain criteria were met.[61] For example, the material must have been of commercial specification grade; a market must have existed for the material; a substantial portion of that type of material must have been made available for the manufacture of a new saleable product; and the material could have been used to replace, or to substitute for, virgin raw material. The person also must demonstrate that he or she exercised "reasonable care" to determine that the receiving facility where the materials were intended to be recycled was in compliance with federal, state, and local environmental laws.

Section 127(f) makes the exemption unavailable if the person claiming the exemption had reason to believe the scrap material would not be recycled by the receiving facility; that the material would be burned as fuel or for energy recovery or incineration; that the receiving facility was not in compliance with federal, state, and local environmental laws; or that hazardous substances had been added to the material; or if the person failed to exercise "reasonable care" in managing and handling the material.[62]

Contributors of "De micromis" and Municipal Solid Wastes

Enacted in the 107th Congress, Section 102(a) of the Small Business Liability Relief and Brownfields Revitalization Act (P.L. 107-118) amended Section 107 of CERLA to limit the liability of parties who contributed only certain quantities or types of wastes to sites listed on the NPL. The amendment did not extend these exemptions to such parties at sites not listed on the NPL. Section 102(a) of P.L. 107-118 added Section 107(o) to CERCLA to exempt from cleanup liability parties who generated or transported waste to a site listed on the NPL, if they contributed only "de micromis" amounts of hazardous substances to that site.[63] To qualify for this exemption, a party must demonstrate that it contributed less than 110 gallons of liquid materials or less than 200 pounds of solid materials containing hazardous substances. The exemption is not available to persons who contributed such quantities of wastes to a site on or after April 1, 2001.

Section 102(a) of P.L. 107-118 also added Section 107(p) to CERCLA to exempt from cleanup liability residential property owners, small businesses, and small non-profit organizations that contributed only municipal solid waste to a site listed on the NPL.[64] The exemption is limited to municipal solid waste generated by a household or that possesses characteristics typical of household waste. The exemption is available only to the generators of the waste and persons who arranged for the transport of the waste. The exemption is not available to the owner or operator of the disposal site where the waste was sent, such as a landfill, nor to transporters of the waste who selected the disposal site.

Bona Fide Prospective Purchasers and Innocent Landowners

Subtitle B of Title II of P.L. 107-118 authorized exemptions from cleanup liability for two categories of parties: (1) "bona fide" prospective purchasers and (2) owners whose properties became contaminated only as a result of migration from a contiguous property owned by another person. Subtitle B also established more specific criteria for the availability of the third party defense to "innocent" landowners who had no knowledge of existing contamination at the time of acquiring a property and had no involvement in the actions that led to the contamination. These exemptions are available to site owners who meet the requisite statutory criteria, regardless of whether the site is listed on the NPL. In contrast, the above exemptions from cleanup liability for contributors of de micromis amounts of wastes and municipal solid wastes are available only at sites listed on the NPL.

Prior to the enactment of P.L. 107-118, EPA had used its existing authorities under Section 122 of CERCLA[65] to enter into voluntary settlement agreements with prospective purchasers who had no involvement in the contamination, as a mechanism to limit their cleanup liability upon acquiring ownership of a contaminated property. EPA commonly referred to these agreements as "Prospective Purchaser Agreements."[66] This type of settlement usually capped a purchaser's financial responsibility for the cleanup, or required less extensive cleanup work by the purchaser. These agreements also typically included a covenant promising that the federal government would not sue for further liability, and offered protection from contribution claims by other liable parties. (See the "Voluntary Settlement Agreements" section of this report for additional discussion.) After the enactment of P.L. 107-118, a "bona fide" prospective purchaser who met the requisite statutory criteria could become eligible for an exemption from cleanup liability without entering into a formal settlement agreement with EPA.

Section 222 of P.L. 107-118 added Section 107(r) to CERCLA, exempting "bona fide" prospective purchasers of contaminated property acquired after the date of the enactment of P.L. 107-118 (January 11, 2002).[67] This exemption is not available to persons who purchased a contaminated property on or before January 11, 2002, and therefore is not retroactive. A person who knowingly purchased a contaminated property on or before that date must have entered into a Prospective Purchaser Agreement with EPA under Section 122 of CERCLA to limit his or her cleanup liability upon acquiring ownership. Section 222 of P.L. 107-118 also added Section 101(40) to CERCLA, defining the term "bona fide prospective purchaser" and specifying the criteria of eligibility for the exemption provided in Section 107(r) of the statute. [68]

In addition to "bona fide" prospective purchasers, Section 221 of P.L. 107-118 added Section 107(q) to CERCLA, exempting owners of contaminated property from cleanup liability if the contamination occurred only as a result of the migration of a hazardous substance from a contiguous property owned by another person. [69] To obtain the exemption, an owner of a contiguous property must have had no knowledge of the presence of the hazardous substance, nor the possibility of its migration, when acquiring ownership. Section 223 of P.L. 107-118 amended the definition of the term "contractual relationship" in Section 101(35) of CERCLA[70] to establish more specific criteria for "innocent" landowners to claim the third party defense against liability under Section 107(b)(3) of CERCLA.[71] Of importance, the exemptions for "bona fide" prospective purchasers and contiguous property owners reference these criteria in the definition of contractual relationship, making the criteria applicable to all three exemptions.

Under these criteria, an owner claiming an exemption as a "bona fide" prospective purchaser, "innocent" landowner, or contiguous property owner must have had no association with the activities that led to the contamination, and must have had no relationship with the persons who caused or contributed to the contamination (aside from a contractual relationship only involving the conveyance of the property). The key difference among these exemptions is that a "bona fide" prospective purchaser may know that a property is contaminated at the time of acquisition, and still be exempt from liability under CERCLA for the cleanup costs. A person claiming an exemption as an "innocent" landowner or a contiguous property owner must prove that he or she had no knowledge, or no reason to know, of the contamination at the time of acquisition.

To demonstrate that a best effort was made to know whether contamination may be present, CERCLA requires a person to have made "all appropriate inquiries" into the previous ownership and uses of the property.[72] A person seeking to claim an exemption from liability as a "bona fide" prospective purchaser, "innocent" landowner, or contiguous property owner must satisfy this requirement *before* acquiring ownership. As directed by P.L. 107-118, EPA promulgated regulations that identify specific measures a person must take to demonstrate that "all appropriate inquiries" were made satisfactorily.[73] The process outlined in the regulations for making "all appropriate inquiries" is similar to a preliminary site assessment, and must be performed by an environmental professional hired at the person's expense.

A person also must satisfy other conditions *after* receiving ownership of a property to maintain an exemption from liability under CERCLA as a "bona fide" prospective purchaser, "innocent" landowner, or contiguous property owner. The owner must take "reasonable steps" to (1) stop any continuing release of a hazardous substance; (2) prevent any future releases; and (3) prevent or limit exposure to any previously released hazardous substance.[74] The owner also must provide any legally required notices of the discovery of hazardous substances on the property, and must comply with any land use restrictions and institutional controls that may be put into place by regulators to prevent potential exposure to the hazardous substances. Satisfying these criteria can have the effect of minimizing, but not eliminating, an owner's responsibility for managing the contamination, even though the owner may be exempt from cleanup liability under CERCLA.

A person who discovered contamination upon conducting "all appropriate inquiries" must take "reasonable steps" to manage the contamination once becoming the owner to maintain exemption status as a "bona fide" prospective purchaser. If a person conducted "all appropriate inquiries" before acquiring a property and still did not discover the contamination, that person must take these steps once the existence of the contamination is known to maintain exemption status as an "innocent" landowner, or a contiguous property owner if the contamination resulted from migration. It should be emphasized that the burden of proof is on the person seeking an exemption from liability to demonstrate that "all appropriate inquiries" were made before acquiring ownership *and* that "reasonable steps" are taken after acquiring ownership to manage the contamination once its existence is known.

HAZARDOUS SUBSTANCE SUPERFUND TRUST FUND

CERCLA established the Hazardous Substance Superfund Trust Fund to provide a source of funds for the federal government to finance the cleanup of contaminated sites where the potentially responsible parties cannot pay or cannot be identified. This assumption of financial responsibility for these "orphan shares" of cleanup costs is intended to ensure that the actions necessary to protect human health and the environment are carried out. The availability of Superfund Trust Fund monies to pay for the cleanup of orphaned sites is subject to appropriations by Congress. Once appropriated, the availability of Superfund monies under EPA's Superfund program to pay for remedial actions is further subject to cost-sharing agreements with the states in which the sites are located, as discussed in the "State Participation" section of this report.

Original Taxing Authority

The special taxing authority to finance the Superfund Trust Fund expired at the end of 1995. Before this authority lapsed, three dedicated taxes on petroleum, chemical feedstocks (and imported chemical derivatives), and corporate income provided most of the revenues for the Superfund Trust Fund. Revenues from the General Fund of the U.S. Treasury also contributed to the trust fund to augment the dedicated taxes, but these general tax revenues were a relatively small portion of the total revenues to the trust fund during the time that the dedicated taxes were collected through the end of 1995.

As originally enacted in 1980, Section 211(a) of CERCLA authorized the Superfund excise taxes on petroleum[75] and chemical feedstocks.[76] Section 515(a) of the Superfund Amendments and Reauthorization Act of 1986 expanded the reach of the tax on domestically manufactured chemical feedstocks to include imported chemical derivatives.[77] Taxing imported derivatives was intended to compensate for the potential loss of revenues as overseas manufacturing of chemical feedstocks increased. Prior to expiration at the end of 1995, the Superfund excise tax on petroleum was 9.7 cents per barrel. The Superfund excise tax on chemical feedstocks and imported chemical derivatives varied from $0.22 per ton to $4.87 per ton, depending on the substance (with the exception of xylene which was taxed at a higher rate of $10.13 per ton in the initial years of the tax until 1992.) Section 516(a) of the Superfund Amendments and Reauthorization Act of 1986 established the special tax on corporate income to provide an additional revenue stream for the Superfund Trust Fund.[78] Prior to expiration in 1995, the Superfund tax on corporate income (formally referred to as the Corporate Environmental Income Tax) was 0.12% of corporate alternative minimum taxable income in excess of $2 million.

Whether to reinstate Superfund taxes has been a long-standing controversy since the taxing authority lapsed at the end of 1995. Congress has considered numerous bills to reauthorize the taxes, but none have been enacted to date. The reauthorization debate has centered around numerous "fairness" issues. Supporters of the taxes maintain that dedicated tax revenues for the Superfund program are necessary to ensure that polluters pay for the cleanup of contamination they have caused or may cause in the future, often referred to as the "polluter pays principle." In this sense, some have characterized Superfund taxes as an "insurance plan" for the public that is intended to provide resources for cleanup in the event that businesses may become bankrupt and cannot be pursued for their liability. On the other hand, opponents of the taxes have observed that not all of the individual businesses subject to the tax may have been involved in activities that resulted in contamination, and that the actual polluters are paying for the cleanup of most Superfund sites through enforcement actions under the liability provisions of

CERCLA. The extent to which Superfund taxes may have affected "innocent" businesses has been a principal question in the debate over the fairness of the tax structure. The Superfund tax on corporate income was intended to raise additional revenues from a wide range of businesses that may have benefitted from the use of hazardous substances in some way. However, this income tax captured all businesses that met the income threshold, regardless of whether a business may have used or disposed of any hazardous substances. Congress created the Superfund taxes on petroleum and chemical feedstocks based on the broadly held assumption that much of the environmental contamination in the United States had been caused as a result of industrial activities that involved these substances. However, not all petroleum and chemical companies may have been involved in actions that led to contamination.

The appropriateness of the Superfund tax on petroleum has been especially controversial in light of the exclusion of petroleum from the cleanup authorities of CERCLA. Because of this exclusion, monies from the Superfund Trust Fund generally have paid for the cleanup of petroleum contamination, only if the contamination includes hazardous substances that are not part of the petroleum product itself. Congress has established other trust funds to address releases of petroleum. Title V of the Superfund Amendments and Reauthorization of 1986 created the Leaking Underground Storage Tank Trust Fund to pay for actions to respond to petroleum released from underground tanks.[79] Title VIII of the Omnibus Budget Reconciliation Act of 1986 (P.L. 99-509) created the Oil Spill Liability Trust Fund to pay for actions to respond to surface releases of petroleum.[80]

Since Superfund taxing authority lapsed at the end of 1995, there have continued to be varying perspectives on how to fund the cleanup of contaminated sites in the most fair manner to ensure that the responsible parties satisfy their liability, while minimizing the financial burden on taxpayers who did not cause or otherwise contribute to the contamination, or who did not benefit in some way from the actions that resulted in the contamination. The current source of revenues for the Superfund Trust Fund is discussed below.

Current Source of Revenues

After the authority to collect the Superfund taxes expired, the remaining revenues from these taxes were expended by the end of FY2003, leaving revenues from the General Fund of the U.S. Treasury as the main source of monies for the Superfund Trust Fund. Although the Superfund taxes have expired, industry has continued to provide some of the funding for the trust fund via corporate income taxes that contribute to the General Fund. (Revenues to the General Fund consist of corporate income taxes, individual income taxes, and miscellaneous federal receipts and collections that are not dedicated to specific federal trust funds.)

In addition to general Treasury revenues, others sources of monies have continued to contribute some revenues to the Superfund Trust Fund for appropriation by Congress. Cleanup costs borne by the federal government that are later recouped from the potentially responsible parties are deposited into the trust fund (referred to as cost recoveries). These recouped funds can be made available for the cleanup of other sites where the potentially responsible parties cannot pay or cannot be found. Fines and penalties assessed against potentially responsible parties for violations of CERCLA are deposited into the trust fund as well. Interest also accrues on the trust fund balance. Collectively, these monies have been relatively small compared to the amount of general Treasury revenues that now support most of the annual discretionary appropriations from the trust fund to implement EPA's Superfund program. However, these other sources of monies do continue to help finance the trust fund, and to some extent reduce the need for general Treasury revenues at sites where the potentially responsible parties cannot be found or cannot pay.

Special Account Funds

Private settlement funds have been an additional source of monies for the Superfund Trust Fund. As amended in 1986, Section 122(b)(3) of CERCLA authorizes EPA to retain funds that it receives from private parties under voluntary settlement agreements to perform the cleanup of sites at which those parties may be liable.[81] (See the "Voluntary Settlement Agreements" section below.) These private settlement funds are deposited into site-specific Special Accounts within the Superfund Trust Fund, which are dedicated to the cleanup of the sites covered under the settlements. EPA has received nearly $4 billion in private settlement funds over time and has deposited these funds into over 1,000 Superfund Special Accounts since the establishment of the first account in FY1990.[82] These funds are available directly to EPA and are not subject to discretionary appropriations by Congress. Once all planned future work at a site is complete, EPA may "reclassify" the remaining balance of a Special

Account for direct obligation to perform cleanup work at other sites, as a means to replace any appropriated funds that also may have been spent at the site covered by the Special Account. In other instances, EPA may transfer the remaining balance of a Special Account to the general portion of the Superfund Trust Fund, which would be subject to subsequent appropriation by Congress.

ENFORCEMENT MECHANISMS

There are three mechanisms through which the federal government can take actions to enforce cleanup liability under CERCLA, if the potentially responsible parties can be identified and have the financial capability to pay. These mechanisms include judicial or administrative orders, cost- recovery actions, and voluntary settlement agreements. Like the response authorities of CERCLA, these enforcement authorities are Presidential authorities. As discussed earlier in this report, a 1987 Executive Order delegated the President's response authorities under CERCLA to EPA and other federal agencies. This order also delegated the enforcement of the statute to EPA at sites on the land, and to the U.S. Coast Guard within inland river ports and harbors, the Great Lakes, and U.S. coastal waters. References in this report to the enforcement authorities of EPA apply equally to the U.S. Coast Guard within its delegated jurisdiction. CERCLA also authorizes citizen suits to enforce the cleanup requirements of CERCLA, but a cleanup action first must be completed before compliance with applicable requirements can be challenged. Each of these enforcement mechanisms is discussed below.

Judicial or Administrative Orders

Section 106(a) of CERCLA authorizes EPA to issue an administrative order, or to pursue a judicial order through the Department of Justice, to require a potentially responsible party to perform cleanup actions to address "an imminent and substantial endangerment to public health or welfare, or the environment" arising from an actual or threatened release of a hazardous substance.[83] Section 106(b)(1) authorizes fines of up to $25,000 per day for failure to comply with a cleanup order.[84] Section 107(c)(3) of CERCLA also allows a party that fails to comply with a cleanup order to be held liable for punitive damages up to three times the costs incurred by the United States out of the Superfund Trust Fund to carry out the cleanup action that the party did not perform.[85] Monies received by the United States for such punitive damages are to be deposited into the trust fund, and can be made available to finance the cleanup of other sites, subject to appropriations by Congress.

If the party who receives and complies with a Section 106 order can prove it is not liable under CERCLA, or that the cleanup actions required by EPA under the order were "arbitrary and capricious" or otherwise not in accordance with law, Section 106(b)(2) authorizes that party to be reimbursed from the Superfund Trust Fund.[86] This provision is intended to protect an innocent party from the costs of enforcement actions that may be imposed inappropriately upon that party, or to prevent a liable party from being required to pay for a more stringent cleanup than may be warranted to protect human health and the environment.

Cost-Recovery Actions

At some sites, EPA may spend Superfund Trust Fund monies upfront to initiate the cleanup if the potentially responsible parties are not yet identified, or if a cleanup order or settlement agreement with the identified parties is not yet finalized. In the event that EPA does expend Superfund monies at a site with viable parties, reimbursement may be included in the terms of any administrative settlement agreement that may be entered into with the parties. EPA also may pursue recovery of Superfund monies from the parties through judicial actions, in conjunction with the Department of Justice.

Section 107(a) of CERCLA specifically authorizes EPA to recover Superfund monies from the potentially responsible parties, as long as those actions are not inconsistent with the NCP.[87] States and Indian tribes, and any other person, who chooses to perform cleanup actions also may recover their costs from the potentially responsible parties, as long as those actions are consistent with the NCP. The costs of health effects studies carried out by the ATSDR under Section 104(i) of CERCLA,[88] and damages for injury to, destruction of, or loss of natural resources (and the assessment of such injury, destruction, or loss), are recoverable as well.

Section 113(g)(2) of CERCLA limits the time during which a cost-recovery action may be commenced against a potentially responsible party, which could reduce a party's financial liability at a site if recovery is not sought quickly enough.[89] Cost-recovery actions must be filed within three years after the completion of a removal action, except for removal actions allowed to extend beyond the general time limit of 12 months. For these lengthier removal actions, the costs can be sought within six years after the determination was made to extend the timing beyond 12 months. Cost-recovery actions must be commenced within six years after the initiation of the physical construction of a remedial action. If the remedial action is initiated within three years after the completion of the removal action that preceded it, the costs of that removal action may be recovered as part of the recovery of the costs of the remedial action that followed.

Voluntary Settlement Agreements

If a potentially responsible party is willing to resolve its liability voluntarily, Section 122 of CERCLA[90] gives EPA the discretion to enter into an administrative settlement agreement with that party instead of pursuing an enforcement action through a judicial or administrative order under Section 106, or a cost-recovery action under Section 107. Avoiding an enforcement action by EPA through a voluntary settlement agreement can save a party the costs of litigation, possibly motivating a party to agree to settle its liability.

Section 122(f) gives EPA the discretion to include a covenant in the agreement promising that the federal government will not sue for further liability.[91] Such a covenant can provide an incentive for a party to agree to perform specific cleanup actions or to make a monetary payment in exchange for a cap on its liability. As discussed in the "Special Account Funds" section above, Section 122(b)(3) authorizes EPA to retain the funds received under a settlement and directly use the funds to fulfill the terms of the settlement.[92] A party who voluntarily settles its liability at a site also is afforded protection from contribution claims by other parties at that site, under Section 113(f)(2) of CERCLA.[93] Such protection is intended to offer yet another incentive for a party to settle, especially if a contribution claim by another party appears imminent.

Whether to enter into a settlement agreement with EPA under Section 122 is entirely voluntary on the part of the potentially responsible party. However, once finalized, the terms of the agreement to perform specific cleanup actions or to make a monetary payment are binding on the party who entered into the agreement. If the party fails to perform the agreed-upon cleanup actions or to pay the agreed-upon costs of the cleanup, Section 109 of CERCLA authorizes civil penalties of up to $25,000 each day that the violation of the agreement continues to occur.[94]

Ability-to-Pay Considerations

In enforcing cleanup liability under CERCLA, EPA has the discretion to consider a potentially responsible party's financial capability in determining that party's share of the cleanup costs. A party with limited financial capability who desires to reduce its share of the cleanup costs may request a reduction in its share through the negotiation of a voluntary settlement agreement with EPA under Section 122 of CERCLA, discussed above. In the negotiation process, the party seeking the reduction must submit financial information to EPA for the agency's consideration to determine whether the party's ability to pay the cleanup costs may in fact be limited.

In 1995, EPA issued its first guidance document on ability-to-pay considerations for use in settlement negotiations. The agency revised its guidance again in 1997.[95] EPA formulated its guidance based on court interpretations of the reach and intent of the cleanup liability provisions of CERCLA,[96] and the agency's policy of balancing two fundamental interests: ensuring that a potentially responsible party satisfies its liability for cleanup, while at the same time not creating an undue financial hardship on that party, or on those who may be dependent upon that party.

In 2002, Congress included provisions in Section 102(b) of Title I of P.L. 107-118 that amended Section 122(g) of CERCLA to establish a new category of *de minimis* settlement that explicitly endorsed EPA's policy to reduce a party's share of the cleanup costs based on that party's ability to pay.[97] Prior to this amendment, CERCLA explicitly authorized *de minimis* settlements only for owners of property who were not involved in the release of hazardous substances and who had no knowledge of any hazardous substances on the property; and for persons who contributed a relatively small amount of hazardous substances that were minimally toxic in comparison to other hazardous substances at the site.

In determining whether a party satisfactorily demonstrates a limited ability to pay, EPA must consider the ability of the person to pay for cleanup actions and "... still maintain its basic business operations, including consideration of the overall financial condition of the person and demonstrable constraints on the ability of the person to raise revenues."[98] Consistent with earlier EPA guidance, the amendment explicitly requires a person seeking a reduced settlement to provide EPA with the financial information that would be necessary to determine the ability of that person to pay for cleanup actions at the site concerned.[99]

If EPA were to grant a reduced settlement, the person who is the subject of the settlement must waive all contribution claims against other potentially responsible parties at the site, unless EPA were to determine that requiring a waiver would be an "unjust" condition.[100] A reduced settlement does not remove a party from the responsibility to provide information and access to the site in the future that may be necessary to carry out the cleanup.[101] After a reduced settlement is finalized, EPA must notify any other potentially responsible parties at the site who have not resolved their liability with the federal government.[102]

Financial capability aside, EPA still may decline a potentially responsible party's request for a reduced settlement, if the agency determines that the party has failed to comply with any request for access, information, or an administrative subpoena in relation to the site, or has impeded or is impeding, through action or inaction, the performance of a cleanup action at the site.[103] If EPA were to determine that a potentially responsible party is not eligible for a reduced settlement, EPA is required to provide the reasons for the determination in writing to the potentially responsible party who requested the reduced settlement.[104]

EPA's determination of a party's eligibility for a reduced settlement is not subject to judicial review,[105] nor is a dispute over an ability-to-pay determination of the agency within the jurisdiction of the EPA Environmental Appeals Board.[106] Consequently, EPA would appear to have final authority to determine a party's ability to pay its share of the cleanup costs, based on the financial information submitted by that party in the settlement negotiation process. In practice, a party only can pay to the extent of its actual financial capability, to the point of bankruptcy. The authority of EPA to reduce a party's share of the cleanup costs is intended to avoid such financial outcomes as a consequence of cleanup liability under CERCLA.

Enforcement Discretion

CERCLA does not require EPA to use any one particular enforcement mechanism at an individual site, but allows the agency enforcement discretion to select which of the above mechanisms would be the most effective in achieving cleanup goals. EPA typically attempts to negotiate voluntary settlement agreements with the potentially responsible parties first, and usually turns to the use of Section 106 orders or Section 107 cost-recovery actions when a negotiated settlement appears unlikely. At a site where there are multiple potentially responsible parties, EPA also has the enforcement discretion to pursue the liability of all, some, or only one party. Even when enforcing against less than all parties, EPA still may recover the full amount of cleanup costs through joint and several liability. As described earlier in this report, joint and several liability means that any liable party can be held responsible for the full cost of cleanup, regardless of the degree of involvement.

EPA usually pursues the liability of parties at a site who are thought to have contributed more greatly to the contamination, and to be more capable of performing or paying for the cleanup. This selective approach is intended to reduce the enforcement transactions costs to the federal government. For the purpose of fairness, Section 113(f)(1) of CERCLA authorizes the parties who are enforced against to recover some of their costs from other potentially responsible parties whom EPA did not elect to pursue.[107] Section 113(f)(3)(B) also authorizes parties who have resolved all (or some) of their liability under settlements with EPA to seek contribution from other parties who are not participants in the settlements.[108] Section 113(f)(2) explicitly protects parties from contribution claims who have entered into settlements with EPA to resolve their liability under Section 122 of CERCLA.[109]

Citizen Suits

Although EPA is responsible for enforcing cleanup liability, Section 206 of the Superfund Amendments and Reauthorization Act of 1986 added Section 310 to CERCLA authorizing citizens to challenge the adequacy of a cleanup action in court.[110] The timing of a citizen suit for these purposes is limited. Section 113(h)(4) of CERCLA does not permit a citizen suit to be brought for violation of a cleanup requirement until the selected cleanup action at a site is completed.[111] Further, a citizen suit may not be brought with regard to a removal action at a site where a remedial action is planned. This limitation on the timing of a citizen suit is intended to allow the complete implementation of cleanup actions planned at a site, prior to subjecting the adequacy of those actions to judicial review to assess their compliance with CERCLA.

Once the cleanup actions are completed, Section 310(a)(1) authorizes a citizen to commence a civil action against any person who is alleged still to be in violation of a "standard, regulation, condition, requirement, or order,"[112] including any provision of a federal facility cleanup agreement issued under Section 120.[113] Section 310(b)(1) requires such suits to be brought in the district court for the district in which the violation is alleged to have occurred.[114] Section 310(c) authorizes the court to require actions to correct the violation and impose civil penalties.[115] Section 310(d)(1) requires the plaintiff to notify the President, the state in which the violation is alleged to have occurred, and the alleged violator 60 days in advance of commencing a civil action,[116] in a manner prescribed by federal regulation.[117] Section 310(d)(2) prohibits citizen suits if the President already has commenced and is "diligently prosecuting" an enforcement action against the potentially responsible party.[118]

Section 310(a)(2) also authorizes a citizen to commence a civil action against the President or any other officer of the United States, including the Administrators of EPA and ATSDR, for alleged failure to perform any non-discretionary act or duty required under CERCLA, including such act or duty required at a federal facility.[119] Section 310(b)(2) requires such suits to be brought in the United States District Court for the District of Columbia.[120] Section 310(c) authorizes the court to order the President or other officer of the United States to perform the act or duty concerned.[121] Section 310(e) requires the plaintiff to notify the Administrator of EPA, or other department or agency, 60 days in advance of commencing a civil action for the alleged failure to perform a non- discretionary act or duty,[122] in a manner prescribed by federal regulation.[123]

Citizen suits are not available for alleged failure of the President or an officer of the United States to perform a non-discretionary act or duty under Section 311 of CERCLA.[124] This provision states that the Secretary of Health and Human Services "shall" establish and support a research and training program to enhance understanding of the potential health risks associated with exposure to hazardous substances. The program also is to research methods and technologies that would detect hazardous substances in the environment and reduce their amount and toxicity. The program is to be carried out through the awarding of grants, cooperative agreements, and contracts, the funding for which is subject to annual appropriations by Congress.

FEDERAL FACILITIES

After CERCLA was enacted in 1980, questions arose as to whether Congress intended federal facilities to be subject to the cleanup authorities and liability provisions of the statute to the same extent as non-federal facilities. As originally enacted, Section 101(21) of CERCLA defined the term "person" for the purposes of the statute to include the federal government,[125] meaning that the reference to persons who may be held liable under Section 107 may include the federal government. However, the original enactment of the law did not otherwise explicitly address the liability of federal agencies, nor the applicability of other provisions of the statute to federal agencies. Section 120 of the Superfund Amendments and Reauthorization Act of 1986 added Section 120 to CERCLA to clarify that federal departments and agencies are subject to the requirements of CERCLA to the same extent as other entities, including the liability and enforcement provisions of the law.[126] To comply with CERCLA, the federal agency with administrative jurisdiction over a federal facility is responsible for performing and paying for the cleanup of contamination out of its own budget, subject to appropriations by Congress.

Section 111(e) of CERCLA explicitly prohibits the use of Superfund Trust Fund monies to clean up federal facilities, as these monies are dedicated to paying for the cleanup of sites where the potentially responsible parties cannot be identified or cannot pay.[127] However, Section 111(e)(3) does allow the use of Superfund Trust Fund monies at an individual federal facility to provide alternative water supplies, if groundwater contamination has migrated beyond the boundary of that facility, and there are other potentially responsible parties connected to that facility in addition to the United States.[128] In all other instances, Superfund Trust Fund monies are not available for the cleanup of federal facilities.

Congress appropriates funding to various accounts of federal agencies to pay for the performance of the cleanup of federal facilities. These funds generally are intended to fulfill the liability of the United States as the owner or operator of these facilities.[129] However, these accounts do not constitute a cleanup liability fund in a broader sense. The funds are authorized to pay for the performance of the cleanup of the federal government's own facilities by federal agencies. However, the funds are not explicitly authorized to pay cleanup cost-recovery or contribution claims that may be submitted to the United States by other parties, either at federal facilities or at non-federal sites where a federal agency may be held liable as a generator or transporter of wastes sent to a site for disposal. The Judgment Fund of the U.S. Treasury has been the source of payments for cleanup claims submitted to the United States to satisfy the federal share of liability under CERCLA, and compromise settlements for such claims.[130] By statute, the Judgment Fund is a permanent, indefinite appropriation that is intended to pay monetary claims against the United States, which are not otherwise provided by Congress through separate appropriations.[131]

EPA and the states play a role in overseeing and enforcing the implementation of CERCLA at federal facilities, although the agencies that administer these facilities actually fund their cleanup. Section 120(e) of the law requires EPA to take the lead in overseeing the cleanup of federal facilities listed on the NPL,[132] but Section 120(f) allows states and local governments to participate in cleanup decisions.[133] The states play a more prominent role in overseeing the cleanup of federal facilities not listed on the NPL. While CERCLA authorizes EPA and the states to oversee the cleanup of federal facilities, certain provisions of the law can limit their ability to direct or dictate how the cleanup process may be carried out.

As discussed below, CERCLA gives EPA decision-making authority to select remedial actions at federal facilities listed on the NPL, but does not explicitly authorize EPA to direct the schedule of performing those actions, nor how those actions are to be operated and maintained over the long- term to ensure their performance. Further, EPA's enforcement of cleanup requirements at federal facilities through court actions is complicated by the limited ability of one federal agency to sue another. With respect to states, CERCLA requires the opportunity to be involved in cleanup decisions, but does not give states any decision-making authority. In practice, these limitations may restrict the extent to which EPA and the states may oversee the cleanup of federal facilities, even though Section 120 of CERCLA specifically requires federal facilities to comply with the requirements of the statute to the same extent as other entities.

Facilities on the National Priorities List

Within 6 months of the listing of a federal facility on the NPL, Section 120(e)(1) of CERCLA requires the federal agency with administrative jurisdiction over the facility to consult with EPA and the appropriate state authorities to begin a Remedial Investigation/Feasibility Study (RI/FS).[134] As discussed earlier in the "Scope of Response Actions" section of this report, an RI/FS involves an investigation of contamination to assess potential risks to human health and the environment, and a study of the feasibility of the remedial alternatives to address those risks. While consultation with EPA and state authorities is required, CERCLA does not give explicit decision-making authority to EPA or the states to dictate precisely how a federal agency performs this investigation and study phase of the cleanup process.

Within 180 days of the completion of the RI/FS and review by EPA, Section 120(e)(2) requires the federal agency with administrative jurisdiction over the facility to enter into an interagency agreement with EPA to govern the remedial actions to be taken at that facility.[135] This agreement provides an opportunity for EPA to formalize how the other federal agency will carry out the cleanup of the facility to satisfy the requirements of CERCLA. Section 120(e)(4) identifies four elements that are to be included in each interagency agreement: (1) a list of the remedial alternatives considered at the facility, (2) identification of the remedial actions selected from among the alternatives, (3) a schedule for completing each remedial action, and (4) arrangements for any long-term operation and maintenance activities that may be necessary to ensure the performance of the remedial actions over time.[136]

If EPA and the federal agency with administrative jurisdiction over the facility cannot agree on the selection of the remedial actions in negotiating an interagency agreement, Section 120(e)(4)(A) authorizes the Administrator of EPA to resolve the dispute and select the remedial actions he or she deems most appropriate to protect human health and the environment.[137] Although the Administrator may delegate this dispute-resolution authority to an officer or employee of EPA, Section 120(g) prohibits the transfer of the Administrator's authorities under Section 120 to any other agency, official, or employee of the United States, by Executive Order of the President or otherwise, or to any other person.[138] This prohibition primarily is intended to ensure that the role of EPA is maintained in determining the selection of remedial actions.

CERCLA does not provide the Administrator of EPA decision-making authority with respect to other elements of an interagency agreement for a federal facility listed on the NPL, namely the schedule for completing the remedial actions and arrangement for any long-term operation and maintenance activities that may be necessary to ensure the performance of those actions over time. These latter elements would appear to be subject to negotiation between EPA and the federal agency with administrative jurisdiction over the facility. If consensus cannot be reached and the agreement finalized within the statutory deadline of 180 days from the completion of the RI/FS, Section 120(e)(5) requires the federal agency with administrative jurisdiction over the facility to report the delay to Congress.[139]

With respect to the timing of the cleanup, Section 120(e)(3) requires the federal agency responsible for the facility to complete the remedial actions "as expeditiously as practicable" once those actions are selected, but does not indicate a specific time frame or deadline for their completion.[140] The timing of a remedial action ultimately depends on the technical feasibility of that action and the availability of appropriations by Congress. Accordingly, Section 120(e)(3) requires federal agencies to notify Congress of the amount of funding needed to carry out the selected remedial actions at their facilities in their annual budget requests.

Notably, the lack of a final interagency agreement governing an entire facility does not preclude individual remedial actions from proceeding to address discrete contaminated sites at a facility. Further, removal actions intended to address more immediate risks are not subject to an interagency agreement. The main reason for this difference is that the time required to finalize an agreement may delay a removal action needed to address an emergency situation. Because of these reasons, some cleanup actions may proceed without an interagency agreement in place, in effect leaving EPA with less formal means to oversee the cleanup.

States and local governments also may play a role in the cleanup of federal facilities listed on the NPL. Section 120(f) of CERCLA authorizes states and local governments to participate in the planning and selection of remedial actions at federal facilities.[141] Participation by states and local governments is to include, but is not limited to, review of all applicable data as it becomes available, and the development of studies, reports, and plans. Section 120(f) specifies that the opportunity for state officials to participate in cleanup decisions at federal facilities is to be provided in accordance with Section 121(f).

As discussed earlier in the "State Participation" section of this report, Section 121(f) requires states to be afforded opportunities for "substantial and meaningful involvement" in initiating, developing, and selecting remedial actions.[142] Section 121(d) also allows state standards to be applied to a remedial action, thereby offering additional opportunity for state participation.[143] However, Section 121(f)(3) specifies that, if a state wishes to challenge a remedial decision at a federal facility, that state must show that the decision is not supported by "substantial evidence" to compel the selection of a different remedy at that facility.[144] Unlike Section 120(f), the participation requirements of Section 121(f) are not extended to local governments.

Facilities Not on the National Priorities List

States play a more prominent role in overseeing the cleanup of federal facilities not listed on the NPL. In acknowledgement of this role, Section 120(a)(4) of CERCLA clarifies the reach of state law at contaminated federal facilities that are not listed on the NPL.[145] This provision stipulates that state cleanup standards or requirements shall apply to a federal facility that is not on the NPL only to the same extent as those standards or requirements would apply to a non-federal site located in that state. In practical terms, a state may not require more stringent cleanup at a federal facility than it would require at a non-federal site possessing comparable characteristics and conditions under which exposure to contamination may occur.

While state cleanup laws generally can be applied to federal facilities not listed on the NPL, CERCLA does not require federal agencies to enter into formal agreements with states to govern cleanup requirements in a fashion similar to interagency agreements with EPA. However, states may have other authorities to identify and enforce cleanup requirements at federal facilities that they oversee. Most notably, federal facilities that store, treat, or dispose of hazardous waste are subject to permits issued by states with federal authority delegated under the Solid Waste Disposal Act.[146] These permits can require "corrective action" to clean up contamination that may result from waste management or disposal practices.[147] In contemplation of such situations, Section 120(i) of CERCLA states that nothing in CERCLA may affect or impair the obligation of federal agencies to comply with requirements of the Solid Waste Disposal Act at the facilities that they administer, specifically including corrective action requirements.[148]

A corrective action that a state may require under the Solid Waste Disposal Act can be similar in scope to a removal or remedial action under CERCLA. This similarity can result in essentially the same stringency of cleanup in practice, regardless of which statute is applied. As such, Solid Waste Disposal Act permits at federal facilities not on the NPL can function much like CERCLA interagency agreements at federal facilities on the NPL, specifying individual actions required to clean up contamination. In this sense, Solid Waste Disposal Act permits can provide a means for a state to formalize and enforce cleanup requirements at many federal facilities that are not listed on the NPL, for which an interagency agreement with EPA is not required.

Transfer of Contaminated Federal Property

Section 120(h) of CERCLA generally requires the United States to clean up contaminated federal property prior to transferring the property out of federal ownership.[149] The policy premise of this provision is that the United States should assume full responsibility for the cleanup of contamination caused by federal activities, and not shift the burden of that responsibility to the recipient merely as a consequence of acquiring the property. Section 120(h) applies to all contaminated federal property declared surplus to the needs of the federal government. The agency with administrative jurisdiction over a surplus federal property usually performs and pays for the cleanup of contamination to fulfill the financial liability of the United States.

As is the case with federal facilities that remain in federal ownership, funds available for the cleanup of surplus federal properties are subject to appropriations by Congress, and are not eligible for Superfund monies. For example, the Department of Defense performs and pays for the cleanup of surplus federal property on closed military installations out of funds appropriated to the Base Realignment and Closure (BRAC) accounts.

Section 120(h) does not bind the United States to cleaning up a surplus federal property for any one particular use. As a result, the reuse of a property is negotiated between the administering federal agency and the recipient of the property. Disagreements over reuse can arise if the recipient intends to use the property for a purpose that would necessitate a level of cleanup that the federal agency may consider too costly, relative to available appropriations to fund the cleanup. The capabilities of cleanup technologies also could constrain the reuse of a surplus federal property, if it would be impractical to achieve a level of cleanup that would be needed to make the property suitable for a use desired by the recipient.

Continuing Liability of the United States

Consistent with the policy premise of Section 120(h) and retroactive liability under Section 107, the United States remains responsible for contamination found not to have been sufficiently remediated after the property is transferred out of federal ownership. Section 120(h)(3) requires the continuing liability of the United States to be specified through a "covenant" incorporated into the deed transferring the property out of federal ownership.[150] The covenant must warrant that all remedial actions necessary to protect human health and the environment have been taken before the date of transfer, and that the United States shall conduct any additional remedial actions found to be necessary after the date of transfer. A clause also must be included in the deed granting the United States access to the property to perform cleanup actions for which it may be responsible.

In practice, the contents of a deed can place certain limitations on the continuing responsibility of the United States. A deed to a transferred federal property typically warrants cleanup only to a level suitable for the land use negotiated prior to transfer. In some cases, a deed may include a restriction prohibiting certain uses that would be considered unsuitable relative to the level of cleanup performed by the United States. Under such deed restrictions, the United States typically assumes responsibility for additional cleanup only to the extent that more work is found to be needed to make the originally agreed-upon use suitable.

If the new owner later wishes to use the property for a different purpose, the new owner typically must assume responsibility for the additional cleanup costs to make the property suitable for that purpose. In some instances, a deed may prohibit certain land uses even if the new owner is willing to pay the cleanup costs. For example, a deed to a decommissioned military training range may prohibit residential or other uses because of the limitations of cleanup technologies to detect and remove unexploded ordnance. Cleanup capabilities may be constrained especially when ordnance is located beneath the surface, or concealed on the surface by dense vegetation.

Transfer of Uncontaminated Parcels

Some surplus federal properties may contain a mix of contaminated and uncontaminated parcels of land. Although the clean parcels may be ready for reuse, the requirement to clean up the contaminated parcels under Section 120(h) of CERCLA prior to transfer could delay the conveyance of the property as a whole. To address such situations, the 102nd Congress enacted the Community Environmental Response Facilitation Act (CERFA; P.L. 102-426) in 1992. This law amended Section 120(h) by adding a new subsection (4) that authorizes the transfer of uncontaminated parcels on a surplus federal property while cleanup continues on the contaminated parcels.[151] This parcel-by-parcel approach is intended to avoid potential delays in the transfer of "clean" surplus federal lands for reuse, especially such lands on closed military installations where economic redevelopment is desired to replace lost jobs. In the event that previously unknown contamination is later discovered after transfer out of federal ownership, Section 120(h)(4)(D) requires that a deed to an uncontaminated parcel still include a covenant warranting that the United States shall conduct any cleanup actions that may become necessary.[152]

Early Transfer of Contaminated Parcels

The cleanup of a contaminated parcel may take several years or more, depending on the type and level of contamination, technical feasibility of cleanup actions, and availability of appropriations to pay for the cleanup. In such situations, the requirement to complete cleanup prior to transfer out of federal ownership could result in delaying the transfer. Enacted in the 104th Congress, Section 334 of the National Defense Authorization Act for FY1997 (P.L. 104-201) amended Section 120(h)(3) of CERCLA to add a new subsection (C) that allows the transfer of a contaminated parcel on a surplus federal property before cleanup is complete, if certain conditions are satisfied.[153] Although Congress enacted this amendment in annual defense authorization legislation, this authority applies to any surplus federal property administered by any federal agency, not just surplus U.S. military property.

Section 120(h)(3)(C) specifically authorizes a deferral of the cleanup covenant to allow the transfer of title to a contaminated parcel on a surplus federal property before cleanup is complete. Federal agencies often refer to this deferral of the covenant as an "early" transfer, although the statute does not use this term. The deed to a contaminated property transferred out of federal ownership must contain assurances that the cleanup still will be carried out after the property leaves federal ownership. The federal agency responsible for the performance of the cleanup also must identify the funding needed to carry out the cleanup in its annual budget requests.

The deed also must restrict the use of the property to purposes that would be protective of human health and environment, while the cleanup proceeds. For example, at the time of transfer, a property may be suitable for industrial use because the risks of exposure to contamination may be within an acceptable range, whereas other uses that would result in potentially harmful exposure would be restricted until the property is cleaned up sufficiently for that purpose. Once cleanup is complete, the United States remains obligated to provide a covenant at that time, warranting that all necessary actions to protect human health and the environment have been taken to make the property suitable for its intended, eventual use.

The early transfer of a contaminated surplus federal property that is listed on the NPL is subject to the concurrence of the Administrator of EPA and the governor of the state in which the facility is located. The early transfer of a contaminated surplus federal property that is not listed on the NPL still requires the concurrence of the governor of the state in which the facility is located, but not EPA. Federal agencies proposing an early transfer also must provide the public at least 30 days advance notice and an opportunity to comment on the proposed transfer before it is executed.

Cleanup Authorities Specific to Military Facilities

Considering that U.S. military facilities constitute a substantial portion of the inventory of contaminated federal facilities, Section 211 of the Superfund Amendments and Reauthorization Act of 1986 required the Secretary of Defense to establish the Defense Environmental Restoration Program to perform the cleanup of U.S. military facilities.[154] This provision also authorized dedicated Defense Environmental Restoration appropriations accounts to fund the program. Section 211 requires the Secretary of Defense to perform the cleanup of U.S. military facilities under the program in accordance with Section 120 of CERCLA, which in turn specifies the applicability of all of the requirements of CERCLA and the liability and enforcement provisions of the law. Section 211 also requires the Secretary to consult with EPA in implementing the Defense Environmental Restoration Program. Notably, the provisions of Section 211 did not amend CERCLA itself, but were treated as "stand-alone" provisions that apply strictly to U.S. military facilities.[155]

The scope of the Defense Environmental Restoration Program includes the performance of the cleanup of military facilities in the United States that *are* or *were* under the jurisdiction of the Secretary and owned by, leased to, or otherwise possessed by the United States at the time the contamination occurred, and as such may include both active and decommissioned military facilities.[156] The inclusion of decommissioned facilities within the program's scope is consistent with retroactive liability under Section 107 of CERCLA, under which the Department of Defense can be held liable for cleanup as the past owner and operator of those facilities. The scope of the program also includes the correction of other environmental damage that may present an imminent and substantial endangerment to the public health, welfare, or the environment (such as the presence of unexploded ordnance on decommissioned military training ranges), and the demolition and removal of unsafe buildings and structures for safety purposes. The scope of the program does not include the payment of cleanup cost-recovery or contribution claims that may be submitted to the United States by other parties to satisfy federal liability arising from activities of the Department of Defense. As discussed above, the Judgment Fund has been the source of federal monies for the payment of such cleanup claims.

The Defense Environmental Restoration Program initially focused on the cleanup of hazardous substances without a consolidated effort in place to address the safety risks of unexploded ordnance on decommissioned military training ranges. In response to concerns among the public about these potential safety hazards, the 107th Congress included provisions in Sections 311[157] and 312[158] of the National Defense Authorization Act for FY2002 (P.L. 107-107) that expanded the scope of the Defense Environmental Restoration Program to include the cleanup of unexploded ordnance, discarded military munitions, and munitions constituents (i.e., hazardous substances leached from munitions into the environment) on decommissioned military training ranges and munitions disposal sites in the United States. The Department of Defense established the Military Munitions Response Program as a sub-element within the Defense Environmental Restoration Program to carry out these requirements.

The statutory authority of the Military Munitions Response Program extends only to decommissioned military training ranges and munitions disposal sites in the United States, but not to operational ranges.[159] Since the enactment of specific cleanup authorities for military facilities in the 1986 amendments to CERCLA, the Department of Defense has expressed long-standing concern that the carrying out of cleanup actions on an operational range could prevent or interrupt its active use for training, and thereby possibly impair military readiness. So far, operational ranges have been subject to federal waste disposal regulations promulgated under the Solid Waste Disposal Act, but not cleanup under CERCLA unless the contamination migrates off-range. EPA promulgated these disposal regulations in 1997, referred to as the Military Munitions Rule.[160] Under this rule, munitions on an operational range are not considered hazardous waste, and therefore are not subject to hazardous waste disposal requirements under the Solid Waste Disposal Act, until they are removed from the range. Upon removal, their disposal is subject to permit requirements for hazardous waste disposal.

Munitions typically are removed from an operational range only to the extent necessary to ensure safe access by military personnel for training purposes. Consequently, much of the munitions may remain on an operational range indefinitely, unless contamination from munitions were to migrate off-range and present a risk of exposure. In such situations, removal of munitions could be pursued to eliminate the source of the contamination. Absent off-range migration, munitions generally may be left on a range as long as the range remains in operational status. In such circumstances, cleanup of the munitions typically is not required until the range is closed, and the range then becomes eligible for cleanup under the Military Munitions Response Program.

National Security Exemption

Although Section 120 of CERCLA clarified the applicability of the statute to federal facilities, Section 120(j) authorized the President to exempt an individual federal facility from a requirement of CERCLA on a case-by-case basis if the exemption would be necessary to protect national security.[161] This exemption is intended to prevent situations in which a federal facility may become unavailable for purposes essential to protecting national security, if carrying out a specific cleanup action somehow may interfere with those purposes. Section 120(j) specifically authorizes the President to exempt an individual facility administered by the Department of Defense or the Department of Energy from compliance with a requirement of CERCLA, if the President deems such an exemption necessary to protect national security.

The President must notify Congress within 30 days of the issuance of an exemption and explain the reason for it. The time period of an exemption initially is limited to one year, but the President may renew it annually with notification to Congress. To date, a national security exemption under CERCLA has not been invoked at any facility of the Department of Defense or the Department of Energy. Instead, contaminated facilities of both departments have been made subject to the cleanup requirements of CERCLA.

Brownfields Properties

In 1993, EPA established an element within the Superfund program to assist communities with the cleanup of certain lower risk sites that did not warrant placement on the NPL, but at which cleanup was desired to encourage economic redevelopment. The purpose of the program was to provide federal financial assistance for the cleanup of properties referred to as "brownfields." These properties typically are abandoned, idled, or underutilized, and on which known or suspected contamination is perceived as a deterrent to redevelopment by prospective purchasers who may be hesitant about becoming liable for cleanup once acquiring ownership.

EPA initially used Superfund appropriations to provide "seed monies" to communities in the form of grants and loans to aid them in financing certain types of cleanup actions. Although there was broad support for this effort, some questioned EPA's authority under CERCLA to use Superfund monies for the cleanup of these lower risk sites that were not listed on the NPL and that did not appear to warrant emergency removal actions under the Superfund program. Still, in the annual appropriations process, Congress set aside funding for brownfields cleanup assistance within the Superfund account for several years without specifically amending CERCLA for this purpose.

In the 107th Congress, Subtitle A and Subtitle C of Title II of the Small Business Liability Relief and Brownfields Revitalization Act of 2002 (P.L. 107-118, hereinafter referred to as the "Brownfields Act") amended CERCLA to provide explicit statutory authority for EPA to administer a Brownfields program separately from the Superfund program. The Brownfields Act authorized appropriations for this new program apart from appropriations for the Superfund account.[162] There had been some concern about the diversion of Superfund appropriations away from addressing the greater human health and environmental risks at NPL sites. Still, the portion of Superfund appropriations that had been spent on the cleanup of brownfields properties was relatively small compared to the total appropriation.

The program explicitly authorized in the Brownfields Act is similar in scope to the program that EPA had established in 1993, with the exception that the Brownfields Act allowed federal financial assistance for the cleanup of contamination resulting from releases of petroleum. As discussed earlier in the "Federal Response Authorities" section of this report, CERCLA otherwise does not apply to the cleanup of petroleum. The Brownfields Act also created two separate types of grants within the Brownfields program. One provides more direct financial assistance for the assessment and cleanup of individual properties. The other provides financial assistance to states and Indian tribes to aid them in carrying out their own cleanup programs, which in turn may assist in the cleanup of individual properties.

Specifically, Section 201 of the Brownfields Act amended Section 104 of CERCLA to add a new subsection (k) that authorized $200 million annually for grants to fund the assessment and cleanup of individual brownfields properties.[163] Entities generally eligible for these grants include state and local governments, Indian Tribes,[164] redevelopment agencies chartered or otherwise sanctioned by a state government, and land clearance authorities or other "quasi-governmental" entities operating under the supervision and control, or as an agent, of a local government. The grants are awarded on a competitive basis. The recipients may use the grant funds to characterize, assess, or remediate brownfields properties, or to capitalize revolving loan funds that in turn may finance the remediation of multiple brownfields properties by other entities, including loans issued to site owners or developers.

Section 231 of the act also added Section 128 to CERCLA, authorizing an additional $50 million annually for other grants to assist states and Indian

Tribes in establishing or enhancing their own cleanup programs.[165] States and Tribes may use these monies to augment their own resources to assist with the cleanup of brownfields properties to prepare them for reuse. They also may use these monies to pursue the cleanup of other contaminated sites within their respective jurisdictions, which may present potential health or environmental risks but are not addressed under the federal Superfund program. The authorization of appropriations for both the Section 104(k) and Section 128 grants expired at the end of FY2006, but Congress has continued to fund these grants through the annual appropriations process without enacting reauthorizing legislation.

As discussed earlier in the "Limitations on Liability" section of this report, Section 222 of the Brownfields Act exempted "bona fide" prospective purchasers of contaminated properties from liability under CERCLA, if they satisfy the prerequisite statutory criteria. This exemption is intended to work in tandem with federal grants assistance under the Brownfields program to further the purpose of stimulating the economic redevelopment of contaminated properties. Accordingly, Section 104(k)(4)(B)(iii) of CERCLA[166] specifically authorizes the eligibility of bona fide prospective purchasers for brownfields grants. If a party cannot qualify for this exemption, or another exemption from liability, that party is not eligible to receive a brownfields grant.[167] This statutory prohibition on awarding Brownfields grants to potentially responsible parties is consistent with the policy premise of the liability scheme of CERCLA to hold the potentially responsible parties responsible for the costs of cleanup, so as to minimize the burden of these costs on the federal taxpayer who had no connection with the site.

Congress also has enacted certain tax incentives to encourage the cleanup of brownfields properties, through amendments to the Internal Revenue Code but not CERCLA itself. These incentives have constituted another form of federal financial assistance to support the cleanup of contaminated sites. Section 941(a) of the Taxpayer Relief Act of 1997 (P.L. 105-34) allowed a taxpayer to fully deduct the costs of cleaning up a brownfields property in the year the costs were incurred.[168] This type of deduction is referred to as "expensing," as opposed to "capitalizing" in which the costs would be deducted over a period of years. The tax incentive was intended to encourage property developers to rehabilitate sites where environmental contamination may be a deterrent to bringing nonproductive properties back into use. The tax deduction has no direct application for public sector entities, such as municipalities, which do not pay income taxes. Enacted in the 111th Congress, Section 745 of the Tax Relief, Unemployment Insurance Reauthorization, and

Job Creation Act of 2010 (P.L. 111-312, Title VII, Subtitle C) extended this brownfields tax incentive through December 31, 2011.

Congress also had authorized another federal brownfields tax incentive in Section 702 of the American Jobs Creation Act of 2004 (P.L. 108-357, Title VII), which expired on December 31, 2009. This incentive addressed the treatment of gain or loss on the sale or exchange of certain qualified brownfields sites,[169] as defined in Section 101(39) of CERCLA.[170] This provision allowed a tax-exempt entity to invest in a qualified brownfields site, and not treat the gains as taxable "unrelated business income." To be eligible for this tax incentive, the entity must have incurred cleanup costs exceeding the greater of $550,000, or 12% of the property's fair market value in a remediated condition, in addition to meeting certain other requirements. The tax incentive was not available to parties who are potentially liable for the cleanup under Section 107 of CERCLA.

End Notes

1 42 U.S.C. § 9604(a).
2 42 U.S.C. § 9615.
3 Executive Order 12580, Superfund Implementation, January 23, 1987, 52 *Federal Register* 2923. Hereinafter, references to Presidential authorities under CERCLA refer to those that have been delegated to EPA and other federal agencies, unless noted otherwise.
4 42 U.S.C. § 9603(a).
5 40 C.F.R. § 302.4.
6 42 U.S.C. § 9601(14).
7 42 U.S.C. § 9601(33).
8 33 U.S.C. § 2701 et. seq.
9 33 U.S.C. § 1321(c).
10 42 U.S.C. § 6991b(h).
11 For EPA's interpretation of the statutory exclusion of releases of petroleum from the cleanup authorities of CERCLA, see Environmental Protection Agency, Office of General Counsel, *Scope of the CERCLA Petroleum Exclusion Under Sections 101(14) and 104(a)(2)*, July 31, 1987. Please note that Section 104(a)(2) of CERCLA, as originally enacted in 1980, defined the term "pollutant or contaminant." Section 101(f) of the Superfund Amendments and Reauthorization Act of 1986 re-designated the definition of this term in Section 101(33) of CERCLA, cited above.
12 42 U.S.C. § 9601(22).
13 40 C.F.R. §§ 302.4, Appendix B.
14 42 U.S.C. § 9607(i).
15 42 U.S.C. § 9607(j).
16 42 U.S.C. § 9604(a)(3).
17 40 C.F.R. Part 300.
18 42 U.S.C. § 9605(a).
19 42 U.S.C. § 9605(c).
20 40 C.F.R. 300.425(c)(3).

[21] For information on the number and status of sites across the United States listed on the NPL over time, see EPA's Superfund program website: http://www.epa.gov/superfund/sites/query/queryhtm/npltotal.htm.
[22] 42 U.S.C. § 9601(23).
[23] 42 U.S.C. § 9601(24).
[24] 40 C.F.R. § 300.425(b)(1).
[25] 42 U.S.C. § 9617.
[26] 42 U.S.C. § 9604(c)(1).
[27] 42 U.S.C. § 9604(c)(3).
[28] 42 U.S.C. § 9604(c)(6).
[29] 42 U.S.C. § 9621(a).
[30] 42 U.S.C. § 9621(b).
[31] 42 U.S.C. § 9621(c).
[32] 42 U.S.C. § 9621(d).
[33] 42 U.S.C. § 9621(d)(4).
[34] 42 U.S.C. § 300f et seq.
[35] 33 U.S.C. § 1313.
[36] 33 U.S.C. § 1314.
[37] 42 U.S.C. § 9621(f).
[38] 42 U.S.C. § 9621(f)(3).
[39] 42 U.S.C. § 9617.
[40] 42 U.S.C. § 9617(e).
[41] 42 U.S.C. § 9604(i).
[42] 42 U.S.C. § 9607.
[43] 42 U.S.C. § 9607(a).
[44] 42 U.S.C. § 9601(8).
[45] 42 U.S.C. § 9601(32).
[46] 33 U.S.C. § 1321.
[47] 42 U.S.C. § 9613(f)(1).
[48] 42 U.S.C. § 9607(f)(1).
[49] 42 U.S.C. § 9607(b).
[50] 42 U.S.C. § 9601(35).
[51] 42 U.S.C. § 9607(i).
[52] 42 U.S.C. § 9607(j).
[53] 42 U.S.C. § 9619.
[54] 42 U.S.C. § 9619(a)(1).
[55] 42 U.S.C. § 9619(a)(2).
[56] 42 U.S.C. § 9619(c).
[57] 42 U.S.C. § 9607(n).
[58] 42 U.S.C. § 9601(20).
[59] 42 U.S.C. § 9627.
[60] 42 U.S.C. § 9627(b).
[61] 42 U.S.C. § 9627(c).
[62] 42 U.S.C. § 9627(f).
[63] 42 U.S.C. § 9607(o).
[64] 42 U.S.C. § 9607(p).
[65] 42 U.S.C. § 9622.
[66] In 1989, EPA developed guidance for entering into Prospective Purchaser Agreements with potentially responsible parties under Section 122 of CERCLA. EPA revised its guidance in 1995 to expand the circumstances under which such agreements would be considered. See Environmental Protection Agency, Office of Enforcement and Compliance Assurance, *Guidance on Agreements with Prospective Purchasers of Contaminated Property*, May 24, 1995.

[67] 42 U.S.C. § 9607(r).
[68] 42 U.S.C. § 9601(40).
[69] 42 U.S.C. § 9607(q).
[70] 42 U.S.C. § 9601(35).
[71] 42 U.S.C. § 9607(b)(3).
[72] 42 U.S.C. § 9601(35)(B).
[73] 40 C.F.R. Part 312.
[74] 42 U.S.C. § 9601(35)(B)(i)(II).
[75] 26 U.S.C. § 4611.
[76] 26 U.S.C. § 4661.
[77] 26 U.S.C. § 4671.
[78] 26 U.S.C. § 59A.
[79] 26 U.S.C. § 9508. For a discussion of the Leaking Underground Storage Tank Trust Fund, see CRS Report RS21201, *Leaking Underground Storage Tanks (USTs): Prevention and Cleanup*, by Mary Tiemann.
[80] 26 U.S.C. § 9509. For a discussion of the Oil Spill Liability Trust Fund, see CRS Report RL33705, *Oil Spills in U.S. Coastal Waters: Background and Governance*, by Jonathan L. Ramseur.
[81] 42 U.S.C. § 9622(b)(3).
[82] For background information on the use and overall status of Superfund Special Accounts, see EPA's Office of Enforcement and Compliance Assurance website: http://www.epa.gov/oecaerth/cleanup/superfund/spec-acct.html.
[83] 42 U.S.C. § 9606(a).
[84] 42 U.S.C. § 9606(b)(1).
[85] 42 U.S.C. § 9607(c)(3).
[86] 42 U.S.C. § 9606(b)(2).
[87] 42 U.S.C. § 9607(a).
[88] 42 U.S.C. § 9604(i).
[89] 42 U.S.C. § 9613(g)(2).
[90] 42 U.S.C. § 9622.
[91] 42 U.S.C. § 9622(f).
[92] 42 U.S.C. § 9622(b)(3).
[93] 42 U.S.C. § 9613(f)(2).
[94] 42 U.S.C. § 9609.
[95] Environmental Protection Agency, Office of Enforcement and Compliance Assurance, *General Policy on Superfund Ability to Pay Determinations*, September 30, 1997. The full text of this document is available on the agency's website: http://www.epa.gov/compliance/resources/policies/cleanup/superfund/genpol-atp-rpt.pdf.
[96] For example, see United States v. Bay Area Battery, 895 F. Supp. 1524 (N.D.Fla. 1995).
[97] 42 U.S.C. § 9622 (g)(7)(A).
[98] 42 U.S.C. § 9622(g)(7)(B).
[99] 42 U.S.C. § 9622(g)(7)(C).
[100] 42 U.S.C. § 9622(g)(8)(A).
[101] 42 U.S.C. § 9622(g)(8)(C).
[102] 42 U.S.C. § 9622(g)(12).
[103] 42 U.S.C. § 9622(g)(8)(B).
[104] 42 U.S.C. § 9622(g)(9).
[105] 42 U.S.C. § 9622(g)(11).
[106] Environmental Protection Agency, Environmental Appeals Board, *The Environmental Appeals Board Practice Manual*, June 2004, "EAB Jurisdiction," pp. 2-4.
[107] 42 U.S.C. § 9613(f)(1).
[108] 42 U.S.C. § 9613(f)(3)(B).
[109] 42 U.S.C. § 9613(f)(2).

[110] 42 U.S.C. § 9659.
[111] 42 U.S.C. § 9613(h)(4).
[112] 42 U.S.C. § 9659(a)(1).
[113] For the purpose of citizen suits, Section 310(a)(1) states that "any person" includes "the United States and any other governmental instrumentality or agency, to the extent permitted by the 11th amendment to the U.S. Constitution."
[114] 42 U.S.C. § 9659(b)(1).
[115] 42 U.S.C. § 9659(c).
[116] 42 U.S.C. § 9659(d)(1).
[117] 40 C.F.R. Part 374.
[118] 42 U.S.C. § 9659(d)(2).
[119] 42 U.S.C. § 9659(a)(2).
[120] 42 U.S.C. § 9659(b)(2).
[121] 42 U.S.C. § 9659(c).
[122] 42 U.S.C. § 9659(e).
[123] 40 C.F.R. Part 374.
[124] 42 U.S.C. § 9659(a), 9660.
[125] 42 U.S.C. § 9601(21).
[126] 42 U.S.C. § 9620.
[127] 42 U.S.C. § 9611(e).
[128] 42 U.S.C. § 9611(e)(3).
[129] The vast majority of the funds are appropriated to the Department of Defense and the Department of Energy for the cleanup of federal facilities which served national defense purposes.
[130] See General Accounting Office, *The Judgment Fund and Litigative Awards Under the Comprehensive Environmental Response, Compensation, and Liability Act*, B-253179, November 29, 1993, available on GAO's website: http://archive.gao.gov/lglpdf63/151167.pdf.
[131] 31 U.S.C. § 1304.
[132] 42 U.S.C. § 9620(e).
[133] 42 U.S.C. § 9620(f).
[134] 42 U.S.C. § 9620(e)(1).
[135] 42 U.S.C. § 9620(e)(2).
[136] 42 U.S.C. § 9620(e)(4).
[137] 42 U.S.C. § 9620(e)(4)(A).
[138] 42 U.S.C. § 9620(g).
[139] 42 U.S.C. § 9620(e)(5).
[140] 42 U.S.C. § 9620(e)(3).
[141] 42 U.S.C. § 9620(f).
[142] 42 U.S.C. § 9621(f).
[143] 42 U.S.C. § 9621(d).
[144] 42 U.S.C. § 9621(f)(3).
[145] 42 U.S.C. § 9620(a)(4).
[146] 42 U.S.C. § 6901 et. seq. The Solid Waste Disposal Act often is referred to as the Resource Conservation and Recovery Act (RCRA; P.L. 94-580), which substantially amended the Solid Waste Disposal Act in 1976 to regulate the storage, treatment, and disposal of hazardous waste.
[147] The Hazardous and Solid Waste Amendments of 1984 (P.L. 98-616) amended the Solid Waste Disposal Act to require operators of hazardous waste facilities to perform corrective actions to clean up environmental contamination resulting from the improper management or disposal of hazardous wastes.
[148] 42 U.S.C. § 9620(i).
[149] 42 U.S.C. § 9620(h).

[150] 42 U.S.C. § 9620(h)(3).
[151] 42 U.S.C. § 9620(h)(4).
[152] 42 U.S.C. § 9620(h)(4)(D).
[153] 42 U.S.C. § 9620(h)(3)(C).
[154] 10 U.S.C. § 2701 et. seq.
[155] Because of its specific applicability to U.S. military facilities, Section 211 of the Superfund Amendments and Reauthorization Act of 1986 is codified in Title 10—"Armed Forces"—of the *United States Code*, rather than Title 42—"Public Health and Welfare"—under which the provisions of CERCLA generally are codified.
[156] 10 U.S.C. § 2701(c).
[157] 10 U.S.C. § 2710.
[158] 10 U.S.C. § 2703(b).
[159] 10 U.S.C. § 2710(d). The statutory scope of the Military Munitions Response Program also specifically excludes any locations outside the United States, the presence of military munitions resulting from combat operations, and operational munitions storage and manufacturing facilities. However, the operation of storage and manufacturing facilities in the United States may be subject to regulation under other statutory authorities.
[160] 40 C.F.R. Part 266, Subpart M.
[161] 42 U.S.C. § 9620(j).
[162] Since the enactment of the Brownfields Act, Congress has appropriated specific levels of funding for Brownfields grants within EPA's State and Tribal Assistance Grants account, and has appropriated funds to administer these grants within EPA's Environmental Programs and Management account.
[163] 42 U.S.C. § 9604(k).
[164] In Alaska, Tribal eligibility is extended specifically to the Alaska National Regional Corporation and Alaska Native Village Corporation as defined in the Alaska Native Claims Settlement Act. The Metakatla Indian Community also is specifically authorized in the statute as being eligible for the grants.
[165] 42 U.S.C. § 9628.
[166] 42 U.S.C. § 9604(k)(4)(B)(iii).
[167] 42 U.S.C. § 9604(k)(4)(B)(i)(IV).
[168] 26 U.S.C. § 198.
[169] 26 U.S.C. § 512(b)(19).
[170] 42 U.S.C. § 9601(39).

In: Oil and Chemical Spills
Editor: Natalie Chrisman
ISBN: 978-1-62948-840-0

Chapter 4

OIL SPILLS IN U.S. COASTAL WATERS: BACKGROUND AND GOVERNANCE*

Jonathan L. Ramseur

SUMMARY

The impacts of an oil spill depend on the size of the spill, the rate of the spill, the type of oil spilled, and the location of the spill. Depending on timing and location, even a relatively minor spill can cause significant harm to individual organisms and entire populations. Oil spills can cause impacts over a range of time scales, from days to years, or even decades for certain spills.

Based on data between 1973 and 2009, the annual number and volume of oil spills have shown declines—in some cases, dramatic declines. However, this trend was altered dramatically by the 2010 *Deepwater Horizon* oil spill in the Gulf of Mexico. The incident led to a significant release of oil: according to the federal government's estimate, the well released approximately 206 million gallons of oil before it was contained on July 15. The 2010 Gulf oil spill generated considerable interest in oil spill governance issues.

---

* This is an edited, reformatted and augmented version of the Congressional Research Service Publication, CRS Report for Congress RL33705, dated January 14, 2013.

This report provides background information regarding oil spills in U.S. coastal waters and identifies the legal authorities governing oil spill prevention, response, and cleanup. The governing framework for oil spills in the United States remains a combination of federal, state, and international authorities. Within this framework, several federal agencies have the authority to implement oil spill regulations. Agency responsibilities can be divided into two categories: (1) oil spill response and cleanup and (2) oil spill prevention/preparedness.

Oil spill response authority is determined by the location of the spill: the U.S. Coast Guard has response authority in the U.S. coastal zone, and the Environmental Protection Agency covers the inland zone. Jurisdiction over oil spill prevention and preparedness duties is determined by the potential sources (e.g., vessels, facilities, pipelines) of oil spills.

As with the *Exxon Valdez* oil spill in 1989, the 2010 *Deepwater Horizon* spill generated significant attention to various oil spill policy matters, including prevention, preparedness, response, and liability and compensation. The 111th Congress enacted three oil spill-related proposals into law (P.L. 111-191, P.L. 111-212, and P.L. 111-281), but these laws generally concerned short-term matters that will not have a lasting impact on oil spill governance.

In general, oil spill-related issues garnered less attention during the 112th Congress. The 112th Congress enacted two statutes that contain oil spill-related provisions. P.L. 112-90 includes several oil spill-related provisions involving pipelines. P.L. 112-141 includes a subtitle referred to as the RESTORE Act. This act directs 80% of any administrative and civil Clean Water Act Section 311 penalties connected with the 2010 *Deepwater Horizon* oil spill into a newly created trust fund. Monies from this fund, through various mechanisms, will support environmental and economic restoration in the Gulf states.

INTRODUCTION

Oil is a dominant source of energy in the United States, supplying the nation with approximately 36% of its energy needs.[1] Its use is widespread, providing fuel for the transportation, industrial, and residential sectors. Vast quantities of oil continuously enter the country via vessel or pipeline and are then transported to destinations throughout the country. With such widespread use and nonstop movement, it is inevitable that some number of spills will occur.

Over the past few decades, several major U.S. oil spills have had lasting repercussions that transcended the local environmental and economic effects. The April 2010 oil spill in the Gulf of Mexico (see text box) has intensified interest in many oil spill-related issues. Prior to the 2010 Gulf spill, the most notable example was the 1989 *Exxon Valdez* spill, which released approximately 11 million gallons (260,000 barrels) of crude oil into Prince William Sound, Alaska. The *Exxon Valdez* spill[2] produced extensive consequences beyond Alaska. According to the National Academies of Science, the *Exxon Valdez* disaster caused "fundamental changes in the way the U.S. public thought about oil, the oil industry, and the transport of petroleum products by tankers ... 'big oil' was suddenly seen as a necessary evil, something to be feared and mistrusted."[3]

Deepwater Horizon Oil Spill in the Gulf of Mexico

On April 20, 2010, an explosion occurred at the *Deepwater Horizon* drilling platform in the Gulf of Mexico, resulting in 11 fatalities. The incident disabled the facility and led to a full evacuation before the platform sank into the Gulf on April 22. A significant release of oil at the sea floor was soon discovered. According to the National Incident Command's Flow Rate Technical Group estimate of August 2, 2010, the well released approximately 206 million gallons of oil (4.9 million barrels) before it was contained.

The following websites provide additional information:

- Federal government's website for the oil spill response and recovery, at http://www.restorethegulf.gov/
- EPA website, at http://www.epa.gov/bpspill/
- NOAA website, at http://www.noaa.gov/deepwaterhorizon/

This report provides background information regarding oil spills[4] in U.S. coastal waters[5] and identifies the legal authorities governing oil spill prevention, response, and cleanup.[6] The first section highlights background issues, including oil spill statistics and potential environmental impacts. The second section discusses the legal and regulatory framework that governs oil spill prevention and response.

BACKGROUND

Oil Inputs

Oil enters coastal waters of the United States from a wide variety of sources. These sources vary considerably. Some sources, such as discharges from recreational vessels, emit relatively minor amounts per individual release but have numerous annual releases, which, in aggregate, contribute a significant annual volume. Other sources, such as spills from oil tankers, release oil on a less frequent basis but have the potential to release a significant volume in one incident. These variances in frequency and volume of oil releases create different environmental impacts as well as different challenges for responders and policymakers.

All Sources

A 2003 National Research Council report groups oil releases into four categories: natural seeps, oil consumption, oil transportation, and oil extraction/production.[7] As illustrated in **Figure 1**, the majority of oil in U.S. waters comes from natural seeps—geologic openings on the ocean floor. Well-known natural seeps are found in the Gulf of Mexico and off the coast of southern California, regions with extensive oil exploration and production. Although the seeps release large volumes of oil each year,[8] the surrounding ecosystem can adapt, and even thrive, because the rate of release is relatively slow.[9]

The vast majority of oil introduced to the environment through human behavior falls into the consumption category. This category is broad in scope and includes land-based sources,[10] operational discharges from commercial vessels[11] and recreational craft,[12] and atmospheric deposition of petroleum hydrocarbons.[13]

The quantitative value and the environmental fate of many of these sources are poorly understood. For example, oil from land-based sources—the largest estimated component of the consumption category—may not directly enter U.S. coastal waters until traveling through various man-made conveyances, such as storm-water drains. As such, the range of uncertainty of land-based runoff is substantial, from a minimum annual estimate of 5.6 million gallons to 588 million gallons.[14]

Potential Sources of Major Spills

Although oil transportation and oil extraction activities contribute (on average) a relatively small percentage of oil to U.S. waters (see **Figure 1**), sources within these sectors have generated major oil spills in U.S. coastal waters. Oil spill policy in the United States has generally focused on prevention, preparation, and response involving oil spills from these (and several other) sources.

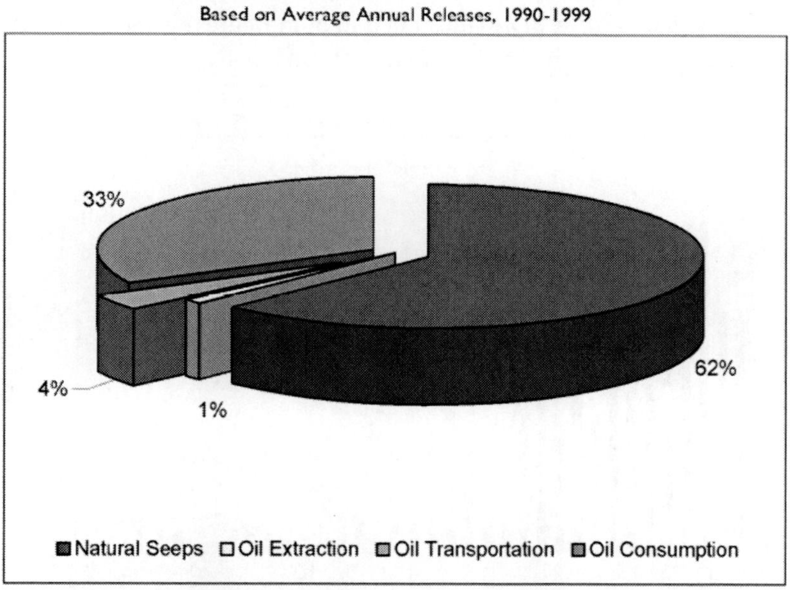

Source: Prepared by the Congressional Research Service (CRS) with data from the National Research Council (NRC) of the National Academies of Science, 2003, *Oil in the Sea III: Inputs, Fates, and Effects*, p. 69.

Notes: Extraction includes platform spills, produced waters, and atmospheric deposition. Transportation includes spills from tanker/barges, pipelines, coastal facilities, and atmospheric deposition. Oil consumption includes river and urban runoff, oil spills from cargo ships, operational discharges from commercial vessels and recreational craft, and atmospheric deposition. For further details of these inputs, see the NRC Report.

Figure 1. Estimates of Percentage Contribution of Oil into North American Coastal Waters, by Major Source Categories.

Figure 2 illustrates the combined number and volume of oil spills from selected sources, whose spills would likely impact U.S. coastal waters. These sources include oil tankers and barges, facilities, and pipelines, among others.[15] Prior to the 2010 Gulf spill, data between 1973 and 2009 indicate that both the number of spills and volume of oil entering U.S. coastal waters have declined; in some years, the declines have been dramatic.[16] Including the 2010 *Deepwater Horizon* oil spill (approximately 200 million gallons) would dramatically alter the volume trendline.

Figure 3 compares the volume of spills over time from the same selected sources identified in **Figure 2**. As **Figure 3** illustrates, the primary source of oil spills in coastal waters has been oil tankers and barges. The substantial drop in the annual *spill volume* (illustrated in both figures) is most attributable to the decline in volume spilled by oil tankers and barges.

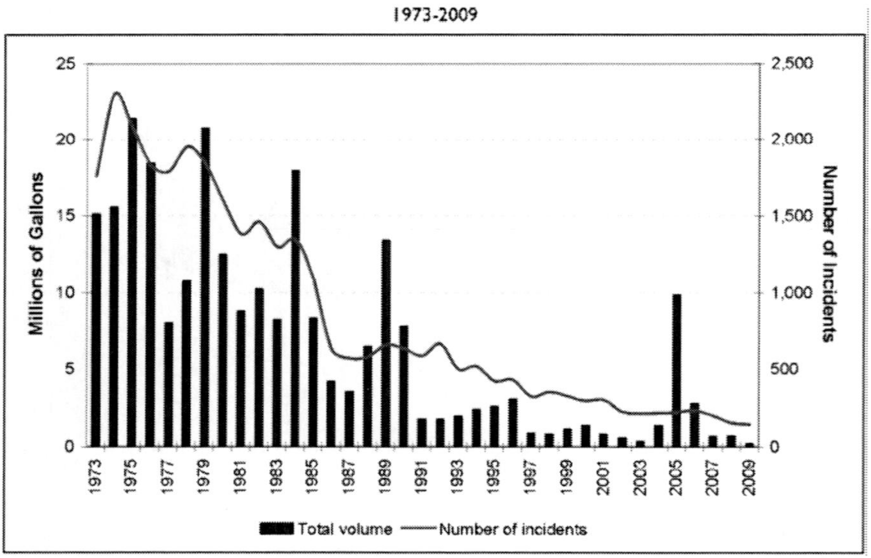

Source: Prepared by CRS with data from the USCG Oil Spill Compendium, at https://homeport.uscg.mil (click on "Investigations"). Accessed in January 2013.

Notes: The Coast Guard states that its Oil Spill Compendium includes spills that have been "investigated" by the Coast Guard. Further, "this data is provided 'as reported,' with no interpretation or filtering. For example, incidents that fall within the jurisdiction of other agencies, or that are not required to be reported under existing Coast Guard regulations, may be included in the compendium."

Figure 2. Annual Volume and Number of Oil Spills from Selected Sources.

1973-2009

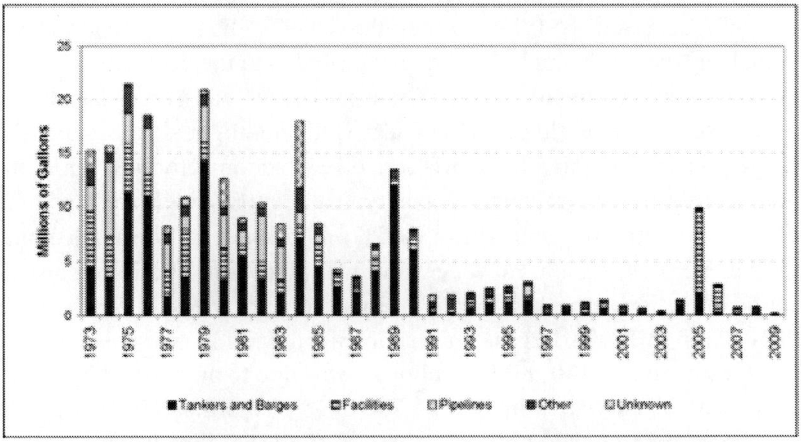

Source: Prepared by CRS with data from the USCG Oil Spill Compendium. Accessed in January 2013.

Notes: The Coast Guard states that its Oil Spill Compendium includes spills that have been "investigated" by the Coast Guard. Further, "this data is provided 'as reported,' with no interpretation or filtering. For example, incidents that fall within the jurisdiction of other agencies, or that are not required to be reported under existing Coast Guard regulations, may be included in the compendium." Other sources include non-tanker vessels, such as cargo ships, passenger vessels, fishing vessels, and recreational vessels. For a complete list see the USCG Oil Spill Compendium at https://homeport.uscg.mil (click on "Investigations").

Figure 3. Comparison of Estimated Oil Spill Volumes from Selected Sources.

The volume of oil spilled from vessels in U.S. waters in the 1990s differed dramatically from the volume spilled in the previous decades. This historical decline of *spill incidents* is likely related, at least in part, to international oil pollution standards that went into effect in 1983. These new standards were implemented in the United States by the Act to Prevent Pollution from Ships (discussed later in this report).[17]

In addition, the *Exxon Valdez* spill of 1989 and the resulting Oil Pollution Act of 1990 (OPA) played key roles in the subsequent spill volume reduction. The 1990 Act (discussed below) made comprehensive changes to U.S. oil pollution law by expanding federal response authority and increasing spill liability. The high costs associated with the *Exxon Valdez* spill,[18] and the threat of broad liability imposed by OPA (in some scenarios, unlimited liability), have likely been the central drivers for the spill volume decline seen in the 1990s. In addition to international and federal governance, 28 states had oil spill liability laws, 19 of which imposed unlimited liability, before the *Exxon Valdez* spill occurred in 1989.[19] After the 1989 spill, some states enacted additional legislation,[20] which may have contributed to the declines.

Although the volume of oil spills from oil tankers and barges has dwarfed other selected sources, the 2010 Gulf spill highlighted the worst-case discharge potential of spills from offshore oil extraction activities. Spills from offshore platforms and pipelines have typically represented (on an annual basis) only a relatively minor (only 0.05%) component of the total input to North American waters.[21] However, oil well blowouts from offshore oil extraction operations have historically been a source of major oil spills. Before the 2010 Gulf spill, the largest unintentional oil spill in world history—the *IXTOC I*, estimated at 140 million gallons—was due to an oil well blowout in Mexican Gulf Coast waters in 1979.[22] The 2010 *Deepwater Horizon* incident released approximately 206 million gallons of oil (4.9 million barrels) before it was contained.[23] As a comparison, the largest oil tanker spill in world history—the *Atlantic Empress* off the coast of Tobago in 1979—was estimated at approximately 84 million gallons.[24]

Impacts of Oil Spills in Aquatic Environments

The impacts of an oil spill depend on the size of the spill, the rate of the spill, the type of oil spilled, and the location of the spill. Depending on timing and location, even a relatively minor spill can cause significant harm to individual organisms and entire populations.[25] Oil spills can cause impacts over a range of time scales, from days to years, or even decades for certain spills. Impacts are typically divided into acute (short-term) and chronic (long-term) effects. Both types are part of a complicated and often controversial equation that is addressed after an oil spill: ecosystem recovery.

Acute Impacts

Depending on the toxicity and concentration of the spill, acute exposure to oil spills can kill various organisms and cause the following debilitating (but not necessarily lethal) effects:[26]

- reduced reproduction,
- altered development,
- impaired feeding mechanisms, and
- decreased defense from disease.

Birds, marine mammals, bottom-dwelling and intertidal species, and organisms in their developmental stages (e.g., fish eggs and larvae) are particularly vulnerable to oil spills.[27]

In addition to the impacts to individual organisms, oil spills can lead to a disruption of the structure and function of the ecosystem. Certain habitats—such as coral reefs, mangrove swamps, and salt marshes—are especially vulnerable, because the physical structure of the habitats depends upon living organisms.

These potential acute effects to individual organisms and marine ecosystems have been "unambiguously established" by laboratory studies and well-studied spills.[28]

Chronic Impacts

Long-term, chronic exposure typically occurs from continuous oil releases—leaking pipelines, offshore production discharges, and non-point sources (e.g., urban runoff). Although spills are normally associated with acute impacts, some oil spills have also demonstrated chronic exposure and effects.[29] There is increasing evidence that chronic, low-level exposures to oil contaminants can significantly affect the survival and reproductive success of marine birds and mammals.[30] However, because of the complexity of factors, including a longer time period and presence of other pollutants, determining the precise effects on species and ecosystems due to chronic oil exposure in a particular locale is difficult for scientists. As a result, studies involving chronic effects are often met with debate and some controversy.

Ecosystem Recovery

Interested parties may have differing opinions as to what constitutes ecosystem recovery. At one end of the spectrum, local groups may demand that an ecosystem be returned to pre-spill conditions. NOAA regulations (15 C.F.R. §990.30) state that recovery "means the return of injured natural resources and services to baseline"—in other words, a return to conditions as they would have been had the spill not occurred. Baseline conditions may not equate with pre-spill conditions. Multiple variables affect local species and ecosystem services. For example, one species at a spill site could have been on the decline at the time of an incident, because of changing water temperatures. These types of trends are considered during the restoration evaluative process (discussed below). Restoration leaves room for site-specific interpretation, which, in the case of the *Exxon Valdez* spill and cleanup, continues to generate considerable argument.

Economic Costs of Oil Spills

The economic costs that can result from an oil spill can be broken into three categories: cleanup expenses, natural resource damages, and the various economic losses incurred by the affected community or individuals.

Cleanup Costs

The cleanup costs of an oil spill can vary greatly and are influenced by a mix of factors: location characteristics, oil type, and oil volume.

Location

Location is generally considered the most important factor because it involves multiple variables. Areas with less water movement, such as marshlands, will generally cost more to clean up than open water. Some spill locations may have relatively robust populations of indigenous micro-organisms that help degrade the oil naturally.[31]

Tourist destinations or sensitive habitats, such as coral reefs, will likely require more stringent cleanup standards, thus increasing the costs. The political and social culture at the spill site plays a part as well. A spill in a high-profile area may receive special attention.[32] Major oil spills, especially ones that affect shoreline ecosystems, are often met with extensive media coverage, placing pressure on parties to take action. Coupled with this pressure, authorities (federal, state, or local) at these locations may require extensive oil spill response requirements, which can influence cleanup cost. For instance, spill costs in the United States are considerably higher than in other parts of the world.[33]

Oil Type

The more persistent and viscous oil types, such as heavy crude oil and intermediates known as bunker fuels, are more expensive to clean up. Gasoline and other lighter refined products may require only minimal cleanup action. Generally, these materials will evaporate or disperse relatively quickly, leaving only a small volume of petroleum product in the environment.

Oil Volume

Compared with other factors, spill volume is less important. A major spill away from shore will likely cost considerably less than a minor spill in a sensitive location. Certainly, the amount of oil spilled affects cleanup costs, because, all things being equal, a larger spill will require a larger and more expensive cleanup effort. However, the relationship between cleanup costs and spill volume is not linear. Cleaning up a smaller spill is likely to cost more than a larger spill on a per-gallon basis.[34]

Natural Resources Damages

This category of costs relates to the environmental impacts caused by an oil spill. Pursuant to OPA, the party responsible for an oil spill is liable for any loss of natural resources (e.g., fish, animals, plants, and their habitats) and the services provided by the resource (e.g., drinking water, recreation).

When a spill occurs, natural resource trustees conduct a natural resource damage assessment to determine the extent of the harm. Trustees may include officials from federal agencies designated by the President, state agencies designated by the relevant governor, and representatives from tribal and foreign governments.[35] The various trustees assess damages to natural resources under their respective jurisdictions.[36] If multiple trustees are involved, they must select a lead administrative trustee (LAT), who coordinates trustee activities and serves as a liaison between oil spill responders. The LAT need not be from a federal agency; however, only a federal LAT can submit a request to the Oil Spill Liability Trust Fund for the initial assessment funding.[37]

The Oil Pollution Act (OPA) of 1990 states that the measure of natural resource damages includes

- the cost of restoring, rehabilitating, replacing, or acquiring the equivalent of the damaged natural resources;
- the diminution in value of those natural resources pending restoration; and
- the reasonable cost of assessing those damages.[38]

Pursuant to OPA, NOAA developed regulations pertaining to natural resource damage assessments in 1996.[39] Natural resource damages may include both losses of direct use and passive uses. Direct use value may derive from recreational (e.g., boating), commercial (e.g., fishing), or cultural or historical uses of the resource. In contrast, a passive-use value may derive from preserving the resource for its own sake or for enjoyment by future generations.[40]

The damages are compensatory, not punitive. Collected damages cannot be placed into the general Treasury revenues of the federal or state government, but must be used to restore or replace lost resources.[41] Indeed,

NOAA's regulations focus on the costs of primary restoration— returning the resource to its baseline condition—and compensatory restoration—addressing interim losses of resources and their services.[42]

Other Economic Costs

Oil spills can generate costs other than response expenses or damages to natural resources. An oil spill can disrupt business activity near the spill, particularly businesses and individuals that count on the resources and reputation of the local environment. For example, the local fishing and tourist industry may be affected. In some cases, a well-publicized oil spill can weaken local or regional industries near the spill site, regardless of the actual threat to human health created by the spill.

Local infrastructure and services can be disrupted by an oil spill. Port and harbor operations may be interrupted, altering the flow of trade goods. Power plants that use cooling water systems may need to temporarily cease operations. For example, the Salem Nuclear Plant—the second-largest nuclear plant in the United States—was forced to halt activity due to a substantial oil spill (more than 250,000 gallons) in the Delaware River in November 2004.

OIL SPILL GOVERNANCE

When the *Exxon Valdez* ran aground in March 1989, there were multiple federal statutes, state statutes, and international conventions that dealt with oil discharges. The spill highlighted the inadequacies of the existing coverage and generated public outrage.[43] Following the spill, Members of Congress faced great pressure to address these issues. (See the **Appendix** for further information concerning these issues.) The end result was the Oil Pollution Act of 1990 (OPA)[44]— the first comprehensive law to specifically address oil pollution to waterways and coastlines of the United States.

The governing framework for oil spills in the United States remains a combination of federal, state, and international authorities. Within this framework, several federal agencies have the authority to implement oil spill regulations. The framework and primary federal funding process used to respond to oil spills are described below.

Oil Pollution Act of 1990

With the enactment of OPA on August 18, 1990, Congress consolidated the existing federal oil spill laws under one program (**Appendix**). The 1990 law expanded the existing liability provisions within the Clean Water Act (CWA)[45] and created new free-standing requirements regarding oil spill prevention and response. Key OPA provisions are discussed below.

Spill Response Authority

When responding to a spill, many considered the lines of responsibility under the pre-OPA regime to be unclear,[46] with too much reliance on spillers to perform proper cleanup.[47] OPA strengthened and clarified the federal government's role in oil spill response and cleanup. OPA Section 4201 amended Section 311(c) of the CWA to provide the President (delegated to the U.S. Coast Guard or EPA) with authority to perform cleanup immediately using federal resources,[48] monitor the response efforts of the spiller, or direct the spiller's cleanup activities. The revised response authorities addressed concerns "that precious time would be lost while waiting for the spiller to marshall its cleanup forces."[49]

The federal government—specifically the On-Scene Coordinator (OSC) for spills in the Coast Guard's jurisdiction—determines the level of cleanup required. Although the federal government must consult with designated trustees of natural resources and the governor of the state affected by the spill, the decision that cleanup is completed and can be ended rests with the federal government. States may require further work, but without the support of federal funding.[50]

National Contingency Plan

The first National Oil and Hazardous Substances Pollution Contingency Plan (NCP) was administratively prepared in 1968 after observing the British government's response to a 37- million-gallon oil tanker spill (*Torrey Canyon*) off the coast of England.[51] The NCP contains the federal government's procedures for responding to oil spills and hazardous substance releases.[52]

OPA expanded the role and breadth of the NCP. The 1990 law established a multi-layered planning and response system to improve preparedness and response to spills in marine environments.[53] Among other things, the act also required the President to establish procedures and standards (as part of the NCP) for responding to worst-case oil spill scenarios.[54]

Tank Vessel and Facility Response Plans

As a component of the enhanced NCP, OPA amended the CWA to require that U.S. tank vessels, offshore facilities, and certain onshore facilities[55] prepare and submit oil spill response plans to the relevant federal agency. In general, vessels and facilities are prohibited from handling, storing, or transporting oil if they do not have a plan approved by (or submitted to) the appropriate agency (discussed below).[56]

The plans should, among other things, identify how the owner or operator of a vessel or facility would respond to a worst-case scenario spill. Congress did not intend for every vessel to have onboard all the personnel and equipment needed to respond to a worst-case spill, but vessels must have a plan and procedures to call upon—typically through a contractual relationship—the necessary equipment and personnel for responding to a worst-case spill.[57]

In 2004, Congress enacted an amendment requiring non-tank vessels (i.e., ships carrying oil for their own fuel use) over 400 gross tons to prepare and submit a vessel response plan.[58] Congress reasoned that many non-tank vessels have as much oil onboard as small tank vessels, thus presenting a comparable risk from an oil spill. Moreover, the international standards for oil spill prevention[59] apply to tanker and non-tanker vessels alike. Thus, the 2004 amendment brought the U.S. law more in line with international provisions.

Double-Hull Design for Vessels

The issue of double hulls received considerable debate for many years prior to OPA, and it was one of the stumbling blocks for unified oil spill legislation. Proponents maintained that double- hull construction provides extra protection if a vessel becomes damaged.[60] However, opponents argued that a double-hulled vessel might cause stability problems if an accident occurred, thus negating the benefits.[61] Stakeholders also highlighted the impacts that a double-hull requirement would entail for the shipping industry (e.g., cost and time of retrofitting, ship availability).[62] The OPA requirements for double hulls reflected some of these concerns.

The act required new vessels carrying oil and operating in U.S. waters to have double hulls.[63] However, OPA provided certain exceptions, depending on the size of the vessel (e.g., less than 5,000 gross tons)[64] and its particular use (e.g., lightering).[65] For older vessels, OPA established a staggered retrofitting schedule, based on vessel age and size. As of January 2010, single-hull vessels (with several exceptions, some of which expire in 2015) cannot operate in U.S. waters.

Liability Issues[66]

OPA unified the liability provisions of existing oil spill statutes, creating a freestanding liability regime. Section 1002 states that responsible parties are liable for any discharge of oil (or threat of discharge) from a vessel or facility[67] to navigable waters, adjoining shorelines, or the exclusive economic zone[68] of the United States (i.e., 200 nautical miles beyond the shore).

Regarding the oil spill statutes prior to OPA, Congress recognized that "there is no comprehensive legislation in place that promptly and adequately compensates those who suffer other types of economic loss as a result of an oil pollution incident."[69] OPA broadened the scope of damages (i.e., costs) for which an oil spiller would be liable. Under OPA, a responsible party is liable for all cleanup costs incurred, not only by a government entity, but also by a private party.[70] In addition to cleanup costs, OPA significantly increased the range of liable damages to include the following:

- injury to natural resources,
- loss of personal property (and resultant economic losses),
- loss of subsistence use of natural resources,
- lost revenues resulting from destruction of property or natural resource injury,
- lost profits and earning capacity resulting from property injury or natural resource injury, and
- costs of providing extra public services during or after spill response.[71]

OPA provided limited defenses from liability: act of God, act of war, and act or omission of certain third parties. These defenses are similar to those of the Superfund statute,[72] established in 1980 for releases of hazardous substances (which does not include oil).

Except for certain behavior, including acts of gross negligence or willful misconduct,[73] OPA set liability limits (or caps) for cleanup costs and other damages. Until 2006, liability limits for vessels were based on vessel carrying capacity, generally $1,200 per gross ton. As an example, the liability limit for the 2004 *Athos* tanker spill in Delaware River was approximately $45 million.[74]

OPA requires the President to issue regulations to adjust the liability limits at least every three years to take into account changes in the consumer price index (CPI). Despite this requirement, adjustments to liability limits were not made until Congress amended OPA in July 2006. The Coast Guard and

Maritime Transportation Act of 2006 (P.L. 109-241) increased limits to $1,900/gross ton for double-hulled vessels and $3,000/gross ton for single-hulled vessels. Furthermore, the Coast Guard made its first CPI adjustment to the liability limits in 2009, increasing the limits to $2,000 and $3,200, respectively.[75]

Mobile offshore drilling units (MODUs), like the *Deepwater Horizon* unit involved in the April 2010 incident in the Gulf of Mexico, are first treated as tank vessels for their liability cap. If removal and damage costs exceed this liability cap, a MODU is deemed to be an offshore facility for the excess amount.[76]

Offshore facility liability is unlimited for removal costs but capped at $75 million for other costs and damages; onshore facility and deepwater port liability is limited to $350 million. In contrast to tank vessel liability limits, these liability limits are at the same level as they were in 1990.

The Oil Spill Liability Trust Fund

Prior to OPA, federal funding for oil spill response was generally considered inadequate,[77] and damages recovery was difficult for private parties.[78] To help address these issues, Congress supplemented OPA's expanded range of covered damages with the Oil Spill Liability Trust Fund (OSLTF).

Pursuant to Executive Order (EO) 12777, the Coast Guard created the National Pollution Funds Center (NPFC) to manage the trust fund in 1991. The fund may be used for several purposes:

- prompt payment of costs for responding to and removing oil spills;
- payment of the costs incurred by the federal and state trustees of natural resources for assessing the injuries to natural resources caused by an oil spill, and developing and implementing the plans to restore or replace the injured natural resources;
- payment of parties' claims for uncompensated removal costs, and for uncompensated damages (e.g., financial losses of fishermen, hotels, and beachfront businesses);
- payment for the net loss of government revenue, and for increased public services by a state or its political subdivisions; and
- payment of federal administrative and operational costs, including research and development, and $25 million per year for the Coast Guard's operating expenses.

Although Congress created the OSLTF in 1986,[79] Congress did not authorize its use or provide its funding until after the *Exxon Valdez* incident. In 1990, OPA provided the statutory authorization necessary to put the fund in motion. Through OPA, Congress transferred balances from other federal liability funds[80] into the OSLTF. In complementary legislation, Congress imposed a 5- cent-per-barrel tax on the oil industry to support the fund.[81] Collection of this fee ceased on December 31, 1994, due to a sunset provision in the law. However, in April 2006, the tax resumed as required by the Energy Policy Act of 2005 (P.L. 109-58). In addition, the Emergency Economic Stabilization Act of 2008 (P.L. 110-343) increased the tax rate to 8 cents through 2016. In 2017, the rate increases to 9 cents. The tax is scheduled to terminate at the end of 2017.[82]

Financial Responsibility

To preserve the trust fund and ensure that responsible parties can be held accountable for oil spill cleanup and damages, OPA requires that vessels and offshore facilities maintain evidence of financial responsibility (e.g., insurance). The Coast Guard's National Pollution Funds Center (NPFC) implements the financial responsibility provisions for vessels; the Bureau of Ocean Energy Management, Regulation, and Enforcement (formerly the Minerals Management Service, MMS) implements this requirement for offshore facilities.

The current levels of financial responsibility are related to the current liability limits for various sources (e.g., vessels, offshore facilities) of potential oil spills. The liability limits differ by potential source. In the case of vessels, whose liability limits are a single dollar amount encompassing both removal costs and other damages, the financial responsibility levels are directly tied to the corresponding liability caps. Current law requires responsible parties for vessels to demonstrate the "maximum amount of liability to which the responsible party could be subjected under [the liability limits in OPA Section 1004; 33 U.S.C. 2704]."

Because the structure of offshore facility liability limit is different than vessels, the corresponding financial responsibility limit provisions differ. Responsible parties for offshore facilities in federal waters must demonstrate $35 million financial responsibility, unless the President determines a greater amount (not to exceed $150 million) is justified (33 U.S.C. 2716(c)). The federal regulations that are authored by this statutory provision (30 C.F.R. Part 254) base the financial responsibility amount—between $35 million and $150 million—on a facility's worst-case discharge volume (as defined in 30 C.F.R.

§253.14). For example, a facility with a worst-case discharge volume over 105,000 barrels[83]—the highest level of worst-case discharge listed in the regulations—must maintain $150 million in financial responsibility.

Other Federal Laws

Although OPA is the primary domestic legislation for oil spills, other federal laws contain provisions that relate to oil spills. Many of these provisions were in place before OPA. The following list is not all-inclusive, but it highlights the main requirements authorized by laws other than OPA.

Clean Water Act

The Clean Water Act (CWA) was the primary federal statute governing oil spills prior to OPA and many provisions continue to apply. A key provision is found in Section 311(b)(3), which prohibits the discharge of oil or hazardous substances into U.S. navigable waters. In addition, the CWA contains various penalty provisions for noncompliance, including violations of the discharge prohibition of Section 311(b).[84]

Pursuant to statutory requirements in the CWA,[85] the EPA crafted regulations[86] for spill prevention control and countermeasure (SPCC) plans in 1973. SPCC plans address the "procedures, methods, and equipment and other requirements for equipment to prevent discharges."[87] The EPA's SPCC plans apply only to non-transportation, onshore facilities that exceed a certain oil storage capacity and that, in the event of a spill, can be reasonably expected, because of their location, to produce an oil discharge that would reach navigable waters or adjoining shorelines of the United States.[88] Unlike other oil spill preparedness provisions, SPCC plans focus more on prevention than on response activities, requiring, for example, secondary containment (e.g., dikes, berms) for oil-storage equipment.

The agency offered several regulatory amendments after the 1973 rulemaking. Following the passage of the Oil Pollution Act of 1990 (OPA),[89] the agency proposed substantial changes and clarifications that were not made final until July 2002. For reasons beyond the scope of this report, the effective date of the 2002 final rule has been extended multiple times; for some parts of the amended rule, the current effective date was January 14, 2009, and for other parts, the effective date was extended to November 10, 2010.[90] However, EPA proposed in July 2010 to extend the date an additional year for most facilities.[91]

Outer Continental Shelf Lands Act

The primary federal law governing oil development and operations in waters in federal jurisdiction is the Outer Continental Shelf Lands Act (OCSLA) of 1953 and its subsequent amendments (43 U.S.C. §§1331-1356). The OCSLA provided the foundation for regulations (30 C.F.R. Parts 250 and 550) that are implemented by the Bureau of Ocean Energy Management (BOEM) and the Bureau of Safety and Environmental Enforcement (BSEE).[92] Sections of these regulations address oil spill prevention and response issues by requiring that various equipment and procedures be in place at offshore facilities.[93]

Pipeline Statutes

The U.S. pipeline network is extensive. One estimate indicates there are more than 33,000 miles of pipelines just in the Gulf of Mexico.[94] Moreover, U.S. inland pipelines are concentrated in coastal areas, particularly in the Gulf states, and these pipelines may have an impact on coastal waters if spills reach waterways that empty into coastal waters.

Several laws govern oil pipelines. The Hazardous Liquid Pipeline Act of 1979 (P.L. 96-129) granted authority to the Department of Transportation (DOT) to regulate various issues regarding oil spills from pipelines. On December 29, 2006, the President signed the Pipeline Safety Improvement Act of 2006 (P.L. 109-468) to improve pipeline safety and security practices, and to reauthorize the federal Office of Pipeline Safety.[95] The Office of Pipeline Safety (OPS), which is part of the DOT, implements provisions concerning pipeline design, construction, operation and maintenance, and spill response planning.[96]

Vessel Statutes

Several federal laws directly or indirectly deal with oil pollution from vessels.[97] Laws concerning navigation reduce the possibilities of vessel collision or hull breach by objects in the waterways.[98] Other laws call for particular vessel design standards. For example, the Ports and Waterways Safety Act of 1972,[99] amended by the Port and Tanker Safety Act of 1978,[100] called for specific construction and equipment design requirements for oil tankers. (As noted, OPA subsequently amended this statute in 1990 to establish a phased-in schedule for double-hulled tankers.) Congress enacted the 1970s legislation to coincide with international initiatives. In fact, many of the federal laws concerning vessel standards and pollution control procedures were written to implement international conventions. These are discussed below.

Recent Pipeline Spills

Kalamazoo River—2010

On July 26, 2010, a pipeline released approximately 800,000 gallons of crude oil of oil into Michigan's Talmadge Creek, a waterway that flows into the Kalamazoo River. As the federal OSC (for the inland zone), EPA established a Unified Command of federal, state and local agencies, and private parties to respond to the spill. Pursuant to the liability provisions in OPA, Enbridge Energy Partners, LLP is the responsible party for the spill.

For more up-to-date information, see EPA's Enbridge oil spill website, at http://www.epa.gov/enbridgespill/index.html.

Yellowstone River—2011

On July 1, 2011, am ExxonMobil pipeline ruptured and released oil into the Yellowstone River near Billings, Montana. EPA leads the federal response activities, coordinating with Montana agencies and other federal agencies. According to the pipeline owner (and cited on EPA's website), the incident discharged an estimated 42,000 gallons.

For more up-to-date information, see EPA's Yellowstone River spill website at http://www.epa.gov/ yellowstoneriverspill/.

Federal Agencies' Responsibilities

The United States shares jurisdiction over its coastal waters with the coastal states. The 1953 Submerged Lands Act (SLA) gave coastal states jurisdiction over the submerged lands, waters, and natural resources (e.g., oil deposits) located, in most cases, within 3 nautical miles off the coastline.[101] The waters, seabed, and natural resources beyond the states' waters are exclusively federal, and extend to the edge of the exclusive economic zone (200 nautical miles from shore). However, the federal government maintains the authority to regulate commerce, navigation, national defense, power production, and international affairs within state waters.

The oil spill legal framework involves implementation by multiple federal agencies. Agency responsibilities can be divided into two categories: (1) oil spill response and cleanup and (2) oil spill prevention/preparedness.

Response

As mentioned above, the National Oil and Hazardous Substances Pollution Contingency Plan (NCP) contains the federal government's framework and operative requirements for responding to an oil spill (and releases of hazardous substances). Although first developed through administrative processes in 1968, subsequent laws have amended the NCP, including the Clean Water Act in 1972; the Comprehensive Environmental Response, Compensation, and Liability Act (CERCLA or Superfund) in 1980; and the Oil Pollution Act (OPA) in 1990. Oil spill response actions required under the regulations of the NCP are binding and enforceable, per these enforcement authorities.

The NCP establishes the National Response System (NRS), a multi-tiered and coordinated national response strategy for addressing oil spills and releases of hazardous substances. The NCP provisions specific to oil spill response are codified in 40 C.F.R. Part 300, Subpart D. Key components of the NRS include the following:

- National Response Team (NRT): composed of representatives from the federal departments and agencies assigned roles in responding to oil spills. The U.S. Coast Guard chairs the NRT when a response is being mounted to a spill in a coastal region.
- Regional Response Teams (RRTs): composed of regional representatives of each NRT member agency, state governments, and local governments. The Coast Guard leads the relevant RRT during responses to oil spills in coastal waters.
- Area Committees (ACs): composed of qualified personnel from federal, state, and local agencies. The primary function of each AC is to prepare an Area Contingency Plan (ACP) for its designated area.
- On-Scene Coordinator (OSC): who directs the response efforts and coordinates all other efforts at the scene.

Oil spill response authority is determined by the location of the spill: the Coast Guard has response authority in the coastal zone, and the EPA covers the inland zone.[102] The OSC has the ultimate authority to ensure that an oil spill is effectively removed and actions are taken to prevent further discharge from the source. The OSC is broadly empowered to direct and coordinate all response and recovery activities of federal, state, local, and private entities (including the responsible party), and will draw on resources available through the appropriate ACPs and RRTs.

Although the OSC must consult with designated trustees of natural resources and the governor of the state affected by the spill, the OSC has the authority and responsibility to determine when removal (i.e., cleanup) is complete.

Other agencies, particularly those on the NRT and relevant RRT, may play a role in response activities. As the chair of the NRT (and vice-chair during oil spills in the coastal zone), EPA may provide response support. For example, during the *Deepwater Horizon* spill response, EPA conducted air and water sampling and provided environmental monitoring support, particularly regarding the use of dispersants.

In addition, NOAA provides scientific analysis and consultation during oil spill response activities.[103] Assistance can include oil spill tracking, cleanup alternatives, and knowledge of at- risk natural resources. Moreover, NOAA experts begin to collect data to assess natural resource damages during response operations.

Prevention and Preparedness

Regarding oil spill prevention and preparedness duties, jurisdiction is determined by the potential sources (e.g., vessels, facilities, pipelines) of oil spills. A series of executive orders (EOs), coupled with memoranda of understanding (MOU), have established the various agency responsibilities.[104] **Table 1** identifies the agencies responsible for implementing prevention and preparedness regulations for the potential sources of oil spills.

Prevention responsibilities include, among other things, assessing whether facilities or vessels have the necessary equipment in place. As discussed above, vessels may be required to have double hulls; facilities may need secondary containment.

Preparedness duties involve oversight tasks, such as evaluating facility and vessel response plans. Preparedness responsibilities also include developing and maintaining contingency plans at various levels: area, regional, and national. Personnel training is a vital component of sustaining readiness. NOAA oil spill experts help train responders in government service and private business.

Table 1. Federal Agency Jurisdiction for Oil Spill Prevention and Preparedness Duties, by Source

Potential Source of Oil Spill	Responsible Agency
Vessels	Coast Guard
Onshore, non-transportation facilities	Environmental Protection Agency
Onshore, transportation facilities	Coast Guard and Department of Transportation
Deepwater ports[a]	Coast Guard and Department of Transportation
Offshore facilities (oil/gas extraction)	Bureau of Ocean Energy Management within the Department of Interior
Offshore pipelines directly associated with oil extraction activities (i.e., "production lines")[b]	Bureau of Ocean Energy Management within the Department of Interior
Offshore pipelines not directly associated with oil extraction activities (i.e., "transmission lines")	Office of Pipeline Safety within the Department of Transportation
Inland pipelines	Office of Pipeline Safety within the Department of Transportation

a. There is only one deepwater port for oil in U.S. coastal waters: the Louisiana Offshore Oil Port (LOOP).

b. For further discussion on federal pipeline jurisdiction, see National Research Council, *Improving the Safety of Marine Pipelines*, National Academies of Science, 1994, pp. 86-89.

In addition, OPA requires agencies to conduct internal examinations to test preparedness.[105] As part of this requirement, the Coast Guard conducts Spills of National Significance (SONS) exercises to analyze the Coast Guard's ability to respond to a major oil spill.

International Conventions

The relationship between international and domestic law can be complex. For example, a "self- executing" agreement taking the form of a treaty, signed by the Executive and ratified with the advice and consent of the Senate, stands on equal footing with federal statute. On the other hand, if an international agreement is not self-executing, implementing legislation may be necessary for the agreement's provisions to be given domestic legal effect, including to provide U.S. agencies with the domestic legal authority necessary to carry out functions contemplated under the agreement. Several federal laws governing oil spills were fashioned to implement obligations contained in international agreements.[106]

International conventions have played an important role in developing consistent standards for oil-carrying vessels from different nations. A primary player in this regard is the International Maritime Organization (IMO), a body of the United Nations, which sets international maritime vessel safety and marine pollution standards. The Coast Guard represents the United States at IMO meetings.

Multiple international conventions concern vessels and their impact on the marine environment. Described below are two selected conventions that contain provisions that are particularly relevant to oil pollution in coastal waters.

MARPOL 73/78

The IMO implements the 1973 International Convention for the Prevention of Pollution from Ships, as modified by the Protocol of 1978 (MARPOL 73/78).[107] Vessels whose nations are signatories to MARPOL are subject to its requirements, regardless of where they sail, and member nations are responsible for the vessels registered under their flag.

MARPOL 73/78 includes six annexes, each covering a different pollution type. Annex I (Prevention of Pollution by Oil) entered into force in 1983[108] and established requirements for controlling oil discharges to sea. Annex I requires vessels to have equipment that minimizes oil discharge, such as oil-water separators, and shipboard oil pollution emergency plans (SOPEPs). Although the SOPEP applicability is similar to that of the vessel response plan (VRP) required by OPA,[109] the purpose of the SOPEP is somewhat different. A SOPEP is intended to provide guidance to the vessel's officers regarding proper onboard emergency procedures when an oil spill occurs,[110] whereas the VRP is more focused on responding to the spill itself.

The United States implements Annex I through the Act to Prevent Pollution from Ships (APPS).[111] APPS applies to all U.S.-flagged ships, irrespective of location, and to all foreign- flagged vessels in U.S. waters or at ports under U.S. jurisdiction. The Coast Guard issues and enforces regulations necessary to carry out the APPS provisions. The Coast Guard inspection program is a key component of its oil spill prevention effort.

Intervention Convention

The 1967 *Torrey Canyon* spill off the coast of Great Britain was one of the first major spills to receive worldwide attention.[112] The incident raised many questions regarding oil spill response, particularly when dealing with vessels from other nations. For example, the incident prompted debate over responses allowable if a nation's waters and environment are threatened by a spill from another nation's vessel. The 1969 International Convention Relating to Intervention on the High Seas in Cases of Oil Pollution Casualties (the Intervention Convention) sought to address these issues.

To implement this convention in the United States, Congress passed the Intervention on the High Seas Act of 1974.[113] Under this act, if the Coast Guard determines there to be a "grave and imminent danger to the coastline or related interests of the United States from pollution or threat of pollution of the sea by convention oil [i.e., as defined in the convention]," the Coast Guard can take action to "prevent, mitigate, or eliminate that danger."

State Laws

As mentioned above, multiple states had oil spill liability laws before the passage of OPA in 1990. During the 15 years prior to OPA's passage, the issue of whether or not to preempt state liability laws was perhaps the primary obstacle to enacting unified oil spill legislation. Proponents of preemption argued that differing state laws—particularly the various levels of liability—frustrate the shipping industry and were contrary to the goal of comprehensive federal legislation.

Preemption opponents maintained that states should be allowed (as with most other federal environmental statutes) to set stiffer standards regarding liability, compensation, and cleanup.[114] In the aftermath of the *Exxon Valdez* spill, the scales tipped to the side of anti-preemption. According to OPA Section 1018 (referred to as a "savings clause"), the act will not preempt any state from imposing "additional liability or requirements" with respect to the discharge of oil or related response activity (e.g., cleanup standards). A 2003 study identified 16 states that impose unlimited liability for oil spills.[115]

There was some concern that the language of OPA's savings clause would allow states to regulate matters typically reserved for the federal government, such as oil tanker construction. To address this issue, the conference report stated that the savings clause would not disturb a 1978 Supreme Court decision that dealt with the intersection of federal and state authority to regulate the shipping industry.[116] In that case, the Court determined that a Washington State law was preempted. The state law had attempted to govern oil tanker design, size, and movement in Puget Sound.[117]

Regardless of the clarification in the conference report, the line between federal and state jurisdiction (i.e., the extent of federal preemption) continues to be tested. In 2000, the Supreme Court struck down (as preempted) a Washington State rule calling for various personnel requirements, such as training, on oil tankers.[118] Similarly, in March 2010, a federal district court in Massachusetts ruled against a state law—finding it preempted—that would affect tanker design, personnel qualifications, and navigation.[119]

CONCLUSION

With the nation a significant producer and consumer of oil, vast quantities are continuously extracted, imported, and transported throughout the United States. Oil is expected to remain a primary source of energy in the United States for at least the next several decades. Future oil spills are inevitable.

As with the *Exxon Valdez* oil spill in 1989, the 2010 *Deepwater Horizon* spill generated significant attention to various oil spill policy matters, including prevention, preparedness, response, and liability and compensation. Members held multiple hearings and introduced numerous bills in the 111th Congress. The 111th Congress enacted three oil spill-related proposals into law (P.L. 111-191, P.L. 111-212, and P.L. 111-281), but these laws generally concerned short-term matters that will not have a lasting impact on oil spill governance.[120]

In general, oil spill-related issues garnered less attention during the 112th Congress. The 112th Congress enacted two statutes that contain oil spill-related provisions.[121]

- On January 3, 2012, the President signed P.L. 112-90 (the Pipeline Safety, Regulatory Certainty, and Job Creation Act of 2011), which included several oil spill-related provisions involving pipelines.
- On July 6, 2012, the President signed P.L. 112-141 (MAP-21), which includes a subtitle referred to as the RESTORE Act. The RESTORE Act establishes the Gulf Coast Restoration Fund in the General Treasury. Eighty percent of any administrative and civil Clean Water Act Section 311 penalties paid by responsible parties in connection with the 2010 *Deepwater Horizon* oil spill will provide the revenues

for the fund. Through several different mechanisms, the fund will support environmental and economic restoration in the Gulf states.

APPENDIX. FEDERAL AUTHORITIES BEFORE AND AFTER THE *EXXON VALDEZ* SPILL

The following list highlights the primary federal authorities that were in effect when the *Exxon Valdez* spill occurred in 1989:

- Clean Water Act (1972): The Clean Water Act (CWA) represented the broadest authority for addressing oil spills at the time of the *Exxon Valdez* spill. Section 311 of the CWA established requirements for oil spill reporting, response, and liability. The act also created a fund (311 Fund), maintained by federal appropriations, that could be used for cleanup and natural resource restoration.
- Deepwater Port Act (1974):[122] This statute addressed oil spills and liability issues at deepwater oil ports. The act also set up the Deepwater Port Fund to provide for prompt cleanup and to compensate damages above liability limits. The fund was financed by a per-gallon tax on oil transferred at a deepwater port.
- Trans-Alaska Pipeline Authorization Act (1973):[123] This act covered oil spills and liability relating to the Trans-Alaska Pipeline System (TAPS). Although the pipeline is constructed over land, spills from it could reach coastal waters via inland rivers. The act created a trust fund, financed through a lessee fee, that could be used to respond to spills and damages from the pipeline.
- Outer Continental Shelf Lands Act Amendments (1978):[124] This act established an oil spill liability structure and rules for oil extraction facilities in federal offshore waters. With this legislation, Congress created the Offshore Pollution Fund, financed by a per-gallon fee on produced oil, that could be used for oil spill cleanup and damages.
- National Oil and Hazardous Substances Pollution Contingency Plan (NCP): The first NCP was administratively prepared in 1968 after observing the British government's response to a 37-million-gallon oil tanker spill (*Torrey Canyon*) off the coast of England.[125] The NCP contains the federal government's procedures for responding to oil spills and hazardous substance releases.[126]

After the *Exxon Valdez* spill, many observers[127] described the above legal collection as an ineffective patchwork. Arguably, each law had perceived shortcomings (discussed below in the context of post-*Exxon Valdez* legislation), and none provided comprehensive oil spill coverage.

For more than 15 years prior to the *Exxon Valdez* incident, Congress made attempts to enact a unified oil pollution law. Several contentious issues produced deadlocks, hindering the passage of legislation. One of the central points of debate, state preemption, dealt with whether a federal oil spill law should limit a state's ability to impose stricter requirements, particularly unlimited liability. Other liability questions also generated debate. For example, if an oil spill occurred, should the owner of the cargo (i.e., oil) be held liable, as was the ship owner/operator? Another point of contention was whether oil-carrying vessels should be required to have double hulls. Although proponents argued that a second hull would help prevent oil spills, the shipping industry raised concern that implementing such a mandate would disrupt oil transportation and potentially affect the national economy. A final issue involved the interaction between domestic legislation (federal and state) and international measures. Some were concerned that if the United States became a party to certain international agreements under consideration in the 1980s,[128] the international standards would preempt federal and state laws, especially those establishing liability limits. Proponents argued that these concerns were overstated and stressed that joining the international agreements was especially important for the United States because of the international nature of oil transportation and associated pollution.

End Notes

[1] Energy Information Administration, *Annual Energy Review*, Primary Energy Consumption Estimates by Source, 1949–2011, September 2012, at http://www.eia.gov/totalenergy/data/annual/#summary.

[2] Note that the *Exxon Valdez* spill ranks only 35th for spill volume on the list of international tanker spills since 1967. See International Tanker Owners Pollution Federation Limited, Historical Data, at http://www.itopf.com/stats.html.

[3] See National Research Council (NRC), *Oil in the Sea III: Inputs, Fates, and Effects*, National Academies of Science (hereinafter "NRC report"), February 2003, p. 11.

[4] In this report, "oil" refers to crude oil and petroleum products, including gasoline and other fuels, unless stated otherwise.

[5] For the purposes of this report, "U.S. coastal waters" is defined broadly to encompass all waters between the shore and the boundary of the U.S. exclusive economic zone (200 nautical miles from shore). Note that in other documents, "coastal" may refer only to state waters, but in this report, the term "coastal waters" includes state and federally regulated waters.

[6] Although oil spills certainly occur in or reach non-coastal U.S. waters, this report focuses on issues and background information related to coastal water spills. However, in many cases, the issues overlap.

[7] NRC report, pp. 67-88.

[8] The NRC estimate for natural seep volume ranges from 24 million to 71 million gallons each year. The "best estimate" (included in **Figure 1**) is 47 million gallons (p. 69).

[9] NRC report, p. 2.

[10] This subcategory is particularly broad: municipal wastewaters, non-refinery industrial discharge, refinery discharges, urban runoff, river discharges, and ocean dumping.

[11] Includes large vessels, such as oil tankers, and smaller vessels, such as fishing boats.

[12] Includes motor boats, jet skis, and other recreational vessels.

[13] Atmospheric deposition generally refers to the process of air pollutants (generated from petroleum combustion) reaching water bodies through various mechanisms (e.g., precipitation). According to the NRC report, "atmospheric deposition supplies hydrocarbons somewhat uniformly to the coastal ocean at relatively low loading rates over large areas" (p. 115).

[14] Based on average, annual releases from 1990-1999. NRC report, pp. 69, 87.

[15] Other sources include non-tanker vessels (e.g., cargo ships, passenger vessels, fishing vessels, and recreational vessels) and unknown sources. For a complete list see the USCG Oil Spill Compendium at https://homeport.uscg.mil (click on "Investigations").

[16] As of the date of this report, the Coast Guard Oil Spill Compendium includes comprehensive data through 2009. This database has been updated periodically, but not on an annual basis.

[17] P.L. 96-478, 33 U.S.C. §1901 et seq. These standards and the U.S. law are discussed later in this report.

[18] The *Exxon Valdez* spill tallied approximately $2 billion in cleanup costs and $1 billion in natural resource damages (not including third-party claims)—in 1990 dollars. Punitive damage claims were litigated for more than 12 years, eventually reaching the U.S. Supreme Court in 2008 (*Exxon Shipping v. Baker*, 128 S. Ct. 2605 (2008)). Plaintiffs were eventually awarded approximately $500 million in punitive damages. An additional $500 million in interest on those damages was subsequently awarded.

[19] CRS Report (out-of-print, available from CRS by request), *Liability Provisions in State Oil Spill Laws: A Brief Summary*, October 1, 1990.

[20] For example, California passed the Lempert-Keene-Seastrand Oil Spill Prevention and Response Act in 1990. More information is available at http://www.dfg.ca.gov/ospr/about/history.html#.

[21] While oil extraction activities contribute approximately 1% of the total oil input to North American waters, the vast majority (95%) of this (1%) oil extraction input comes from operational discharges, which are regulated by a Clean Water Act permit system. NRC Report, Table 3-2.

[22] NRC report, p. 33.

[23] National Incident Command's Flow Rate Technical Group estimate of August 2, 2010.

[24] For a list of the largest oil tanker spills, see The International Tanker Owners Pollution Federation (ITOPF) website, at http://www.itopf.com/.

[25] NRC report, p. 4.

[26] These "sub-lethal" effects can occur at concentrations that are several orders of magnitude lower than concentrations that cause death. NRC report, p. 127.

[27] NRC report, Chapter 5; also multiple conversations with National Oceanic and Atmospheric Administration (NOAA) personnel (2008).

[28] NRC report, p. 120.

[29] NRC report, p. 121.

[30] NRC report, p. 134.

[31] See, for example, Terry Hazen et al., "Deep-Sea Plume Enriches Indigenous Oil-Degrading Bacteria," *Science* (Online), August 24, 2010; Richard Camilli et al., "Tracking

Hydrocarbon Plume Transport and Biodegradation at *Deepwater Horizon*," *Science* (Online), August 19, 2010.

[32] For example, the November 7, 2007, spill (53,000 gallons) from a container ship into the San Francisco Bay generated considerable interest.

[33] The average cleanup cost is three times higher in the United States than in Europe (based on 1997 data and *excluding* the Exxon Valdez costs). See Etkin, Dagmar, "Estimating Cleanup Costs for Oil Spills," paper presented at the 1999 International Oil Spill Conference, 1999, citing data from the Oil Spill Intelligence Report International Oil Spill Database.

[34] This is primarily due to the fact that a spill of any size (e.g., in a sensitive area) will require that equipment and response experts be sent to the scene. See Etkin, Dagmar, "Estimating Cleanup Costs for Oil Spills," paper presented at the 1999 International Oil Spill Conference, 1999, p. 5.

[35] For more information, see NOAA's Damage Assessment, Remediation, and Restoration Program at http://www.darrp.noaa.gov/about/index.html.

[36] 33 U.S.C. §2706(c). In some cases, trustees may share responsibility over the same resource. See, for example, Department of the Interior's "Pollution Response and Natural Resource Trusteeship Training Module On NRDA," at http://www.doi.gov/oepc/response/a01.htm.

[37] 33 U.S.C. §2712 and Executive Order (EO) 12777 (October 18, 1991).

[38] 33 U.S.C. §2706(d).

[39] 61 *Federal Register* 440 (January 5, 1996). See also NOAA, *Injury Assessment Guidance Document for Natural Resource Damage Assessment Under the Oil Pollution Act of 1990* (1996).

[40] See 15 C.F.R. §990.30, definition of "value."

[41] 33 U.S.C. §2706(f); William D. Brighton, *Natural Resource Damages under the Comprehensive Environmental Response, Compensation, and Liability Act* (2006), U.S. Department of Justice, Environment and Natural Resources Division.

[42] William D. Brighton, *Natural Resource Damages under the Comprehensive Environmental Response, Compensation, and Liability Act* (2006), U.S. Department of Justice, Environment and Natural Resources Division.

[43] A handful of other oil spills followed the *Exxon Valdez* in 1989 and 1990 (e.g., the *Mega Borg* spilled 5 million gallons of oil in the Gulf of Mexico), further spurring congressional action.

[44] P.L. 101-380, primarily codified at U.S.C. §2701 et seq.

[45] The official statutory name is the Federal Water Pollution Control Act, P.L. 92-500, as amended, codified at 33 U.S.C. §1251 et seq.

[46] See, for example, Wilkinson, Cynthia et al., "Slick Work: An Analysis of the Oil Pollution Act of 1990," *Journal of Energy, Natural Resources, and Environmental Law*, 12 (1992), p. 190.

[47] See Grumbles, Benjamin, and Manley, Joan, "The Oil Pollution Act of 1990: Legislation in the Wake of a Crisis," *Natural Resources and Environment*, 10:2 (1995), p. 38.

[48] Leading up to the passage of OPA, parties referred to this approach as "federalizing" the spill.

[49] U.S. Congress, House Committee on Merchant Marine and Fisheries, Report accompanying H.R. 1465, Oil Pollution Prevention, Removal, Liability, and Compensation Act of 1989, 1989, H.Rept. 101-242, Part 2, 101st Cong., 1st sess., p. 84.

[50] OPA §1011.

[51] See EPA "National Contingency Plan Overview" at http://www.epa.gov/emergencies/content/lawsregs/ncpover.htm.

[52] The NCP is codified at 40 C.F.R. Part 300.

[53] OPA §4202, amending §311(j) of the CWA.

[54] OPA §4201(b), amending §311(d)(2)(J) of the CWA.

[55] The response plan requirement is applicable only to an onshore facility that, because of its location, could reasonably be expected to cause substantial harm to the environment by discharging into navigable waters, adjoining shorelines, or the exclusive economic zone. CWA §311(j)(5)(iii).

[56] OPA §4202, amending §311(j)(5)(E) of the CWA.

[57] U.S. Congress, House Committee on Merchant Marine and Fisheries, Report accompanying H.R. 1465, Oil Pollution Prevention, Removal, Liability, and Compensation Act of 1989, 1989, H.Rept. 101-242, Part 2, 101st Cong., 1st sess., p. 87. OPA §4202, amending §311(j)(5)(C)(iii) of the CWA.

[58] Amendments Relating to the Oil Pollution Act of 1990, Title VII of Coast Guard and Maritime Transportation Act of 2004 (P.L. 108-293), codified at 33 U.S.C. §1321.

[59] Primarily the shipboard oil pollution emergency plans required by MARPOL 73/78, discussed later in this report.

[60] A study from the National Academy of Sciences reached this conclusion in 1999. See National Research Council, *Double hull Tanker Legislation: An Assessment of the Oil Pollution Act of 1990*, National Academies of Science, 1999, p. 144.

[61] Opponents maintained that if water entered the space between hulls, the ship could become unstable, hindering salvage and possibly capsizing. Cynthia Wilkinson et al., "Slick Work: An Analysis of the Oil Pollution Act of 1990," *Journal of Energy, Natural Resources, and Environmental Law*, 12 (1992), p. 196.

[62] U.S. Congress, Conference Report accompanying H.R. 1465, Oil Pollution Act of 1990, H. Conf. Rept. 101-653, at 140-141 (1990).

[63] OPA §4115, amending 46 U.S.C. §3703.

[64] This exception applied to many inland barges.

[65] Lightering is the process of transferring oil from a large vessel to a smaller vessel. This common practice occurs in designated areas that are typically many miles away from shore.

[66] For a discussion of liability issues raised by the 2010 *Deepwater Horizon* oil spill, see CRS Report R41679, *Liability and Compensation Issues Raised by the 2010 Gulf Oil Spill*, by Jonathan L. Ramseur.

[67] The definition of "facility" is broadly worded and includes pipelines and motor vehicles. OPA §1001.

[68] Under the pre-OPA regime (primarily the CWA), a discharge 12 miles beyond shore had to affect the natural resources before liability attached. Under OPA §1002, the discharge itself triggers liability. Cynthia Wilkinson et al., "Slick Work: An Analysis of the Oil Pollution Act of 1990," *Journal of Energy, Natural Resources, and Environmental Law*, 12 (1992), p. 201.

[69] U.S. Congress, House Committee on Merchant Marine and Fisheries, Report accompanying H.R. 1465, Oil Pollution Prevention, Removal, Liability, and Compensation Act of 1989, 1989, H.Rept. 101-242, Part 2, 101st Cong., 1st sess., p. 31.

[70] OPA §1002(b)(1).

[71] OPA §1002(b)(2).

[72] Section 107(b) of the Comprehensive Environmental Response, Compensation, and Liability Act (CERCLA, commonly known as Superfund), P.L. 96-510.

[73] In addition, liability limits are unavailable if the violation of a federal safety, construction, or operating requirement proximately caused the spill. Spillers must also report the incident and cooperate with response officials to take advantage of the liability caps. OPA §1004(c).

[74] 37,895 gross tons x $1,200/ton = $45.47 million. Vessel data from United States Coast Guard, *Investigation into the Striking of Submerged Objects by the Tank Vessel Athos I in the Delaware River on November 26, 2004 with a Major Discharge of Oil*, January 2006, p. 4.

[75] U.S. Coast Guard, "Consumer Price Index Adjustments of Oil Pollution Act of 1990 Limits of Liability—Vessels and Deepwater Ports," *Federal Register* Volume 74, No. 125 (July 1, 2009), pp. 31357-31369.

[76] 33 U.S.C. 2704(b).

[77] Wilkinson, Cynthia et al., "Slick Work: An Analysis of the Oil Pollution Act of 1990," *Journal of Energy, Natural Resources, and Environmental Law*, 12 (1992), p. 188.

[78] U.S. Congress, House Committee on Merchant Marine and Fisheries, Report accompanying H.R. 1465, Oil Pollution Prevention, Removal, Liability, and Compensation Act of 1989, 1989, H.Rept. 101-242, Part 2, 101st Cong., 1st sess., p. 35.
[79] Omnibus Budget Reconciliation Act of 1986 (P.L. 99-509).
[80] The CWA §311(k) revolving fund; the Deepwater Port Liability Fund; the Trans-Alaska Pipeline Liability Fund; and the Offshore Oil Pollution Compensation Fund.
[81] Omnibus Budget Reconciliation Act of 1989 (P.L. 101-239). Other revenue sources for the fund include interest on the fund, cost recovery from the parties responsible for the spills, and any fines or civil penalties collected.
[82] Section 405 of P.L. 110-343.
[83] This amount is significantly less than the 4.9 million barrels estimated to have been released during the 2010 Gulf spill. See National Incident Command's Flow Rate Technical Group, press release, August 2, 2010.
[84] For further discussion, see CRS Report R41370, *Federal Civil and Criminal Penalties Possibly Applicable to Parties Responsible for the Gulf of Mexico Oil Spill*, by Robert Meltz.
[85] Section 311(j)(1) of the 1972 CWA called for regulations to prevent the discharge of oil from vessels, onshore facilities, and offshore facilities. Executive Order 11735 (August 3, 1973) granted EPA the authority to regulate non- transportation-related onshore and offshore facilities.
[86] U.S. EPA, "Oil Pollution Prevention: Non-Transportation Related Onshore and Offshore Facilities," *Federal Register*, vol. 38, no. 237 (December 11, 1973), pp. 34164-34170.
[87] CWA §311(j)(1)(C).
[88] See 40 C.F.R. §112.1.
[89] P.L. 101-380, primarily codified at U.S.C. §2701 et seq.
[90] For a comprehensive history of the regulations, see *Federal Register*, vol. 74, pp. 58784 (November 13, 2009).
[91] For more information, see EPA's SPCC website at http://www.epa.gov/emergencies/content/spcc/index.htm.
[92] These agencies replaced the former Minerals Management Service (MMS). On May 19, 2010, the Secretary of the Department of the Interior (DOI) replaced the MMS with the Bureau of Ocean Energy Management, Regulation, and Enforcement (BOEMRE). On October 1, 2011, DOI divided BOEMRE into three separate entities: the Bureau of Ocean Energy Management (BOEM), the Bureau of Safety and Environmental Enforcement (BSEE), and the Office of Natural Resources Revenue (ONRR).
[93] For more information, see CRS Report RL33404, *Offshore Oil and Gas Development: Legal Framework*, by Adam Vann.
[94] See, for example, MMS Press Release from February 2, 2005, at http://www.mms.gov/ooc/press/2005/ press0202.htm.
[95] See 49 U.S.C. §60101 et seq.
[96] For further information on pipeline legislation, see CRS Report R41536, *Keeping America's Pipelines Safe and Secure: Key Issues for Congress*, by Paul W. Parfomak.
[97] For a comprehensive list of federal maritime legislation see USCG, *Marine Safety Manual*, Vol. IX (undated), Chapter 1, available at http://homeport.uscg.mil.
[98] For example, the Rivers and Harbors Act of 1899, as amended (33 U.S.C. §401 et seq.), and the International Regulations for Preventing Collisions at Sea, as amended (33 U.S.C. §1601 et seq.).
[99] P.L. 92-340, 33 U.S.C. §1221 et seq.
[100] P.L. 95-474, codified at 33 U.S.C. §§1221-1232 and 46 U.S.C. §§3701-3718.
[101] Most state waters extend 3 nautical miles (1 nautical mile = 6,076 feet, or 1.15 miles) from shore. Louisiana waters extend 3 imperial nautical miles (1 imperial nautical mile = 6,080 feet). Texas and Gulf Coast of Florida waters extend 3 marine leagues (equating to 9 nautical miles). See the MMS, OCS, website ("Definitions and Jurisdictions") at

http://www.mms.gov/incidents/pollution.htm. See also CRS Report RL33404, *Offshore Oil and Gas Development: Legal Framework*, by Adam Vann.

102 The terms inland zone and coastal zone are defined in the National Contingency Plan (40 C.F.R. §300.5). The coastal zone covers all waters subject to the tide, the Great Lakes, and all seaward waters (extending 200 nautical miles beyond shore). The inland zone covers all other U.S. waters. Spills in inland waters can potentially affect coastal waters and ecosystems, particularly if the spill occurs in water systems near the coast. In fact, a fine line may separate specific inland and coastal waters (e.g., consider the nexus between a bay and a river).

103 For more information see NOAA's Office of Response and Restoration website, at http://response.restoration.noaa.gov/index.php.

104 Executive Order (EO) 12777 (October 18, 1991) delegates authorities pursuant to the Oil Pollution Act of 1990. This order was amended by EO 13286 (March 5, 2003), which reorganized duties in response to the creation of the Department of Homeland Security.

105 As required by OPA §4202(a), which amended CWA §311(j)(7), codified in 33 U.S.C. §1321(j)(7).

106 If a treaty is considered "self-executing," domestic legislation implementing the treaty is not necessary. For more details on these issues, see CRS Report RL32528, *International Law and Agreements: Their Effect Upon U.S. Law*, by Michael John Garcia.

107 For convention texts and other materials, see http://www.imo.org.

108 The phrase "entry into force" signifies that the requisite number of nations have ratified the convention or annex, thus making the agreed upon requirements binding for all participating nations. For more discussion of the procedures of international conventions, see the IMO website at http://www.imo.org.

109 All vessels of any type over 400 gross tons traveling over international waters must have a SOPEP approved by their flag state. See USCG VRP/SOPEP "FAQs" at http://www.uscg.mil/vrp.

110 USCG, 1997, *Marine Safety Manual, Marine Environment Protection, Volume IX*, p. 4-24.

111 P.L. 96-478, 33 U.S.C. §1901 et seq.

112 The *Torrey Canyon*, a Liberian-flagged tanker, spilled approximately 35 million gallons of crude oil.

113 P.L. 93-248 , 33 U.S.C. §1471 et seq.

114 One argument against preemption was that existing requirements under particular state laws would be diminished or negated entirely. See Benjamin Grumbles and Joan Manley, "The Oil Pollution Act of 1990: Legislation in the Wake of a Crisis," *Natural Resources and Environment*, 10:2 (1995), p. 38.

115 Dagmar Etkin, 2003, *A Worldwide Review of Marine Oil Spill Fines and Penalties*, at http://www.environmental- research.com/erc_papers/ERC_paper_10.pdf. See also, CRS Congressional Distribution Memorandum, "Oil Spill Liability in the Gulf States," July 2, 2010 (on file with author).

116 U.S. Congress, Conference Report accompanying H.R. 1465, Oil Pollution Act of 1990, H. Conf. Rept. 101-653, at 122 (1990).

117 *Ray v. Atlantic Richfield*, 435 U.S. 151 (1978).

118 *United States v. Locke*, 529 U.S. 89 (2000).

119 *United States v. Massachusetts*, 2010 Westlaw 1345018 (D. Mass. March 31, 2010).

120 See CRS Report R41453, *Oil Spill Legislation in the 111th Congress*, by Jonathan L. Ramseur.

121 See CRS Report R41684, *Enacted and Proposed Oil Spill Legislation in the 112th Congress*, by Jonathan L. Ramseur.

122 P.L. 93-627, codified at 33 U.S.C. §1501 et seq.

123 P.L. 93-153, codified at 43 U.S.C. §1651 et seq.

124 P.L. 95-372, codified at 43 U.S.C. §1801 et seq.

125 See EPA "National Contingency Plan Overview" at http://www.epa.gov/emergencies/content/lawsregs/ncpover.htm.

[126] The NCP is codified at 40 C.F.R. Part 300.

[127] See, for example, U.S. Congress, House Committee on Merchant Marine and Fisheries, Report accompanying H.R. 1465, Oil Pollution Prevention, Removal, Liability, and Compensation Act of 1989, 1989, H.Rept. 101-242, Part 2, 101st Cong., 1st sess., p. 32.

[128] The two agreements under consideration were the 1984 Protocols to the International Convention on Civil Liability for Oil Pollution Damage and the Protocols to the International Fund for Compensation for Oil Pollution Damages.

In: Oil and Chemical Spills
Editor: Natalie Chrisman
ISBN: 978-1-62948-840-0

Chapter 5

OIL POLLUTION ACT OF 1990 (OPA): LIABILITY OF RESPONSIBLE PARTIES*

James E. Nichols

SUMMARY

The Oil Pollution Act of 1990 (OPA) establishes a framework that addresses the liability of responsible parties in connection with the discharge of oil into the navigable waters of the United States, adjoining shorelines, or the exclusive economic zone. Among other provisions, OPA limits certain liabilities of a responsible party in connection with discharges of oil into such areas. The liability limitations established by OPA are currently the subject of significant congressional interest in the wake of the Deepwater Horizon oil spill in the Gulf of Mexico.

A responsible party is strictly liable for removal costs plus damages resulting from an oil spill incident under OPA. A responsible party's liability under OPA is confined to specific categories of damages. Pursuant to OPA, the total liability for damages in connection with an oil spill is limited based on the type of vessel or facility involved, and the amount of oil discharged. OPA provides limited defenses which, if applicable, allow a responsible party to discharge its liability to persons injured by a discharge of oil into the navigable waters of the United States, adjoining shorelines, or the exclusive economic zone.

* This is an edited, reformatted and augmented version of a Congressional Research Service report, R41266, dated June 2, 2010.

Claims for removal costs and certain damages must, with limited exception, be presented directly to the responsible party. In the event that a claim for removal costs or certain damages is not paid by the responsible party within 90 days, a claimant may present such a claim directly to the Oil Spill Liability Trust Fund or file suit in court. OPA and its regulations establish procedures for recovering removal costs and damages against the Oil Spill Liability Trust Fund.

This report addresses liability under OPA for removal costs and damages, and the basic procedure for recovering removal costs and damages from the Oil Spill Liability Trust Fund in the event that the responsible party fails to settle such claims.

BACKGROUND

The Oil Pollution Act of 1990 (OPA) establishes a framework that addresses the liability of responsible parties in connection with the discharge of oil into the navigable waters of the United States, adjoining shorelines, or the exclusive economic zone.[1] Among other provisions, OPA limits certain liabilities of a responsible party in connection with discharges of oil into such areas.[2]

OPA replaced the liability limitations established in Clean Water Act section 311[3] with much higher ones and expanded the class of persons authorized to recover removal costs from the responsible party to "any person" who has incurred removal costs in connection with a discharge of oil into covered waters.[4] Under the Clean Water Act, only the federal government could recover removal costs from the responsible party. Additionally, OPA created categories of damages for which the responsible party would be liable, within specified limits, based on the type of vessel or facility involved in the oil spill incident.[5] The following sections of this report will discuss liability under OPA for removal costs and damages, and the basic procedure for recovering removal costs and damages from the Oil Spill Liability Trust Fund[6] in the event that the responsible party fails to pay such claims.

LIABILITY OF RESPONSIBLE PARTIES[7]

Under OPA, a responsible party is strictly and jointly and severally liable for removal costs plus damages in connection with a discharge of oil into covered waters.[8] A responsible party's liability for damages, however, is limited under OPA.[9] The Oil Pollution Act of 1990 also states that additional liability may be imposed upon a responsible party under state law.[10]

Responsible Party Liability for Removal Costs

A responsible party in an oil spill incident is liable for removal costs[11] under OPA:

> Notwithstanding any other provision or rule of law, and subject to the provisions of [OPA], each responsible party for a vessel or a facility from which oil is discharged, or which poses the substantial threat of a discharge of oil, into or upon the navigable waters or adjoining shorelines or the exclusive economic zone is liable for the *removal costs* and damages ... that result from such incident.[12]

Removal costs may be recovered from a responsible party by the United States, affected states and Indian tribes, and by any person, to the extent that such person has undertaken removal actions pursuant to the National Contingency Plan, mandated by Clean Water Act section 311.[13] In general, claims for removal costs must be presented first to a responsible party.[14] If the party to whom the claim is presented denies all liability for the claim, or if the claim is not settled by payment within 90 days after the claim was presented, the claimant may elect either to initiate an action in court[15] against the responsible party or to present the claim directly to the Oil Spill Liability Trust Fund.[16]

In limited situations, however, certain claims for removal costs may be presented initially to the Oil Spill Liability Trust Fund.[17] For example, if the President (acting through the Director of the National Pollution Funds Center) advertises or otherwise notifies claimants in writing, then such claimants may bypass the responsible party and present claims for removal costs directly to the Fund.[18] A responsible party may also present claims for removal costs directly to the Fund.[19] The Governor of an affected state may present a claim directly to the Fund for removal costs incurred by that state.[20] Finally, a "United States claimant" may present a claim for removal costs in the event that a foreign offshore unit has discharged oil causing damage for which the Fund is otherwise liable.[21]

In the event that a person presents a claim for removal costs to the Fund, in addition to the aforementioned procedural requirements the claimant must establish (1) that the actions taken were necessary to prevent, minimize, or mitigate the effects of the oil spill incident; (2) that the removal costs were incurred as a result of the actions taken to prevent, minimize, or mitigate the effects of the oil spill incident; and (3) that the actions taken were determined by the Federal On-Scene Coordinator to be consistent with the National Contingency Plan, or were otherwise directed by the Federal On-Scene Coordinator.[22] The amount of compensation for removal cost claims against the Fund is "the total of uncompensated reasonable removal costs of actions taken that were determined by the [Federal On-Scene Coordinator] to be consistent with the National Contingency Plan or were directed by the [Federal On-Scene Coordinator]."[23]

Removal activities for which costs are being claimed must be coordinated by the Federal On-Scene Coordinator, except in "exceptional circumstances."[24] Accordingly, costs incurred for removal activities that are not coordinated with, or directed by, the Federal On-Scene Coordinator may not be recoverable against the Fund.[25]

Responsible Party Liability for Damages

Under OPA, responsible parties are liable for certain damages resulting from an oil spill incident.[26]

> Notwithstanding any other provision or rule of law, and subject to the provisions of [OPA], each responsible party for a vessel or a facility from which oil is discharged, or which poses the substantial threat of a discharge of oil, into or upon the navigable waters or adjoining shorelines or the exclusive economic zone is liable for the removal costs and *damages* ... that result from such incident.[27]

Generally, claims for damages must be presented first to a responsible party.[28] If the party to whom the claim is presented denies all liability for the claim, or if the claim is not settled by payment within 90 days after which the claim was presented, the claimant may elect either to initiate an action in court against the responsible party, or to present the claim directly to the Oil Spill Liability Trust Fund.[29]

As discussed below, some categories of damages are available for any person affected by the oil spill incident, while other categories of damages are only recoverable by the United States, states, and/or political subdivisions of states. Unlike removal costs, which are also determined on a per-incident and per-responsible party basis but are uncapped, damages are capped under OPA on a per-incident and per-responsible party basis, unless certain exceptions apply.[30] In the event that an exception applies to a particular oil spill incident, a responsible party's liability for damages under OPA is unlimited.[31] Although the act does not expressly prohibit the recovery of punitive damages against the responsible party, courts that have considered the issue have held that punitive damages are not recoverable against a responsible party under the act.[32]

Limits on Liability

Limitations on liability for damages under OPA are determined on a per-responsible party and per-incident basis.[33] The act further limits a responsible party's liability for damages based on the type of vessel or facility from which the discharge of oil flows. Specifically, the act provides different liability limits for (1) tank vessels; (2) vessels, generally; (3) offshore facilities (other than deepwater ports); (4) onshore facilities and deepwater ports; and (5) mobile offshore drilling units.[34]

For example, the Deepwater Horizon rig, which exploded in the Gulf of Mexico on April 20, 2010, and sank two days later, is classified as a mobile offshore drilling unit.[35] Initially, a mobile offshore drilling unit is deemed to be a tank vessel for the purposes of determining the responsible party's liability.[36] In the event that damages exceed the liability limits for discharges from tank vessels, the mobile offshore drilling unit is deemed to be an offshore facility.[37] Thus, if damages in the Gulf of Mexico exceed the liability limit established for tank vessels under OPA,[38] the Deepwater Horizon would be deemed an offshore facility, and the liability for the "responsible party" could be limited to removal costs plus $75 million.[39]

The limitations on liability mentioned above, however, are subject to several exceptions. Specifically, if the Deepwater Horizon oil spill incident is determined to be the result of gross negligence, willful misconduct, or violation of federal safety, construction, or operating regulations by a responsible party, then the relevant liability limitation is not applicable.[40] Additionally, the liability limitations established by the act are not applicable as to the responsible party if the responsible party fails or refuses to (a) report the incident (as required by law), (b) cooperate with federal removal activities, or (c) comply with the National Contingency Plan.[41]

Natural Resources

Responsible parties are liable to the United States, states, Indian tribes, or foreign governments for the harm to natural resources caused by an oil spill incident under OPA.[42] The term "natural resources" is broadly defined by the act to include "land, fish, wildlife, biota, air, water, ground water, drinking water supplies, and other such resources belonging to, managed by, held in trust by, appertaining to, or otherwise controlled by" the United States (including the resources of the exclusive economic zone), any state or local government, any Indian tribe, or any foreign government.[43]

Although the term "natural resources" is defined broadly, damages for "injury to, destruction of, loss of, or loss of use of, natural resources, including the reasonable costs of assessing the damages" are only recoverable by trustees acting on behalf of the United States, states, Indian tribes, or foreign governments.[44]

Generally, claims for natural resource damages must be presented directly to the responsible party. If not paid by the responsible party, OPA and its implementing regulations establish a procedure through which uncompensated natural resource damages may be recovered by affected claimants from the Oil Spill Liability Trust Fund.[45]

Real or Personal Property

Responsible parties are liable to claimants under OPA for harm to real or personal property resulting from an oil spill incident.[46] Specifically, a claimant who owns or leases an affected property may recover damages from the responsible party "for injury to, or economic losses resulting from destruction of, real or personal property."[47]

In the event that the responsible party fails to settle a claim for damages to real or personal property within 90 days, a claimant may present such a claim directly to the Oil Spill Liability Trust Fund. To recover damages against the Fund for "injury" to real or personal property, a claimant must prove (1) property ownership or control; (2) property damage or injury; (3) the cost of repairing or replacing the property; and (4) the property values before and after the oil spill incident.[48]

To recover damages against the Fund for "economic loss resulting from the destruction" of real or personal property, in addition to the aforementioned requirements listed above, a claimant must prove (1) that the affected property was not available for use, and if it had been available for use, the value of that use; (2) the availability of substitute property, and if used in lieu of the injured property, the costs associated with such use; and (3) a nexus between the economic losses claimed and the injury to the property.[49]

Loss of Subsistence Use

Under OPA, responsible parties are liable to claimants for the "loss of subsistence use of natural resources."[50] Any claimant who "uses natural resources [for subsistence purposes] which have been injured, destroyed, or lost" as a result of an oil spill incident may recover damages from the responsible party.[51]

In the event that a responsible party fails to settle a claim for damages for loss of subsistence use of natural resources within 90 days, a claimant may present such a claim directly to the Oil Spill Liability Trust Fund. To recover against the Fund, a claimant must (1) identify each specific natural resource for which compensation for loss of subsistence use is claimed; (2) describe the actual subsistence use made of each natural resource; (3) describe how and to what extent the subsistence use was affected by the oil spill incident; (4) describe the mitigation efforts undertaken by the claimant; and (5) describe alternative sources or means of subsistence available to the claimant during the period of time for which loss of subsistence is claimed, and any compensation available to the claimant for loss of subsistence.

Government Revenues

Under OPA, responsible parties are liable for damages "equal to the net loss of taxes, royalties, rents, fees, or net profit shares due to the injury, destruction, or loss of real property, personal property, or natural resources."[52] Only the United States, states, and political subdivisions of states may recover damages from the responsible party for lost government revenues.[53] Accordingly, private entities may not recover damages, either against the responsible party or against the Oil Spill Liability Trust Fund, under this section of the act.

In the event that a responsible party fails to settle a claim for damages for lost government revenues within 90 days, a claimant may present such a claim directly to the Oil Spill Liability Trust Fund. To recover against the Fund, a claimant must (1) identify and describe the economic loss for which compensation is claimed; (2) prove a causal nexus between the loss of revenue and the destruction of real or personal property, or natural resources; (3) provide the total assessment or revenue collected for comparable revenue periods; and (4) establish the net loss of revenue.[54]

The total compensation allowable for lost government revenue claims presented directly to the Fund is the total net revenue actually lost as a result of the oil spill incident.[55]

Profits and Earning Capacity

Damages "equal to the loss of profits or impairment of earning capacity due to the injury, destruction, or loss of real property, personal property, or natural resources" are recoverable by any claimant against the responsible party under OPA.[56] As mentioned above, the act broadly defines the term "natural resources" to include land, fish, wildlife, biota, air, water, ground water, drinking water supplies, and "other such resources."[57]

In the event that a responsible party fails to settle a claim for damages for lost profits and earning capacity within 90 days, a claimant may present such a claim directly to the Oil Spill Liability Trust Fund. To recover against the Fund, a claimant must prove (1) that real or personal property or natural resources have been injured, destroyed, or lost; (2) that the claimant's income was reduced as a consequence of the injury, and the amount of the reduction; (3) the amount of profits or earnings in comparable periods as compared to the period when the claimed loss was suffered; and (4) whether alternative employment or business was available and undertaken.[58]

The amount of compensation allowable for "lost profits and earning capacity" claims presented directly to the Fund is limited to the actual net reduction or loss of earnings or profits suffered.[59]

Public Services

In the event that an affected state or political subdivision of an affected state incurs costs related to the provision of "increased or additional public services during or after removal activities, including protection from fire, safety, or health hazards, caused by a discharge of oil," the responsible party is liable for such costs.[60] Under OPA, only states and their political subdivisions may recover the costs associated with providing increased or additional public services (to the extent that such states are affected by the discharge of oil).[61] Accordingly, the United States is precluded from recovering costs related to the provision of public services from the responsible party under the act.

In the event that a responsible party fails to settle a claim for damages related to the provision of increased or additional public services within 90 days, a claimant may present such a claim directly to the Oil Spill Liability Trust Fund. To recover against the Fund, a claimant must establish (1) the specific type of the public services provided and the need for such services; (2)

that the public services occurred during removal activities; (3) that the services were provided as a result of the oil spill incident and would have not otherwise been provided; and (4) the total costs for the provision of public services and the methods used to compute such costs.[62]

The total amount of compensation allowable for claims presented directly to the Fund for the increased or additional provision of public services is the net cost of the increased or additional service provided by the affected state or political subdivision of the affected state.[63]

Defenses to Liability

OPA provides limited defenses to liability. A responsible party is not liable for removal costs or the damages mentioned above if it can establish—by a preponderance of the evidence—that the discharge or substantial threat of discharge of oil and the resulting damages or removal costs were caused solely by

> (1) an act of God; (2) an act of war; (3) an act or omission of a third party, other than an employee or agent of the responsible party or a third party whose act or omission occurs in connection with any contractual relationship with the responsible party (except where the sole contractual arrangement arises in connection with carriage by a common carrier by rail), if the responsible party establishes, by a preponderance of the evidence, that the responsible party—(A) exercised due care with respect to the oil concerned, taking into consideration the characteristics of the oil and in light of all relevant facts and circumstances; and (B) took precautions against foreseeable acts or omissions of any such third party and the foreseeable consequences of those acts or omissions; or (4) any combination of paragraphs (1), (2), (3).[64]

In other words, OPA requires a responsible party to pay for removal costs plus certain damages resulting from the oil spill, unless the incident was caused by a serious and unanticipated naturally occurring event (e.g., earthquake, hurricane, tornado, etc.), an act of war, an act or omission by a third party with no contractual relationship to the responsible party, or any combination of the aforementioned circumstances. For the "act or omission of a third party" defense to be available to the responsible party, the responsible party must have been operating its facility or vessel in a non-negligent manner.[65]

Pursuant to OPA, these defenses to liability are not available to a responsible party who fails or refuses (a) to report the oil spill incident as required by law, (b) to provide reasonable cooperation and assistance with removal activities, or (c) to comply, without sufficient cause, with the President's general removal authority.[66] Additionally, if a responsible party has actual knowledge of a discharge or a substantial threat of a discharge of oil, it cannot escape liability for removal costs or damages by transferring ownership of the facility (or property upon which the facility is located) to an innocent third party.[67]

OIL SPILL LIABILITY TRUST FUND

Although the Oil Spill Liability Trust Fund is not the focus of this report, the purpose and operation of the Fund should be briefly explained. The Oil Spill Liability Trust Fund is a federally administered trust fund that may be used to pay costs related to federal and state oil spill removal activities; costs incurred by federal, state, and Indian tribe trustees for natural resource damage assessments; and unpaid damages claims.[68] The Fund is financed by a per-barrel tax on crude oil received at United States refineries, and on petroleum products imported into the United States for consumption.[69] The maximum amount of money that may be withdrawn from the Fund is $1 billion per incident.[70] Currently, the Fund may not receive advances from the United States Treasury, as its authority to borrow expired December 31, 1994.[71] The United States Attorney General, however, may commence an action on behalf of the Fund, against a responsible party, to recover any money paid by the Fund to any claimant pursuant to OPA.[72]

APPENDIX A. OIL POLLUTION ACT OF 1990: DEFINITIONS

The Oil Pollution Act of 1990 defines the key terms used throughout the act. For the reader's convenience, this appendix provides definitions of key terms as they appear in 33 U.S.C. § 2701.

Definitions

For the purposes of this Act, the term—

(1) "**act of God**" means an unanticipated grave natural disaster or other natural phenomenon of an exceptional, inevitable, and irresistible character the effects of which could not have been prevented or avoided by the exercise of due care or foresight;

(2) "**barrel**" means 42 United States gallons at 60 degrees Fahrenheit;

(3) "**claim**" means a request, made in writing for a sum certain, for compensation for damages or removal costs resulting from an incident;

(4) "**claimant**" means any person or government who presents a claim for compensation under this title;

(5) "**damages**" means damages specified in section 1002(b) of this Act [33 U.S.C. § 2702(b)], and includes the cost of assessing these damages;

(6) "**deepwater port**" is a facility licensed under the Deepwater Port Act of 1974 (33 U.S.C. § 1501-1524);

(7) "**discharge**" means any emission (other than natural seepage), intentional or unintentional, and includes, but is not limited to, spilling, leaking, pumping, pouring, emitting, emptying, or dumping;

(8) "**exclusive economic zone**" means the zone established by Presidential Proclamation Numbered 5030, dated March 10, 1983, including the ocean waters of the areas referred to as "eastern special areas" in Article 3(1) of the Agreement between the United States of America and the Union of Soviet Socialist Republics on the Maritime Boundary, signed June 1, 1990;

(9) "**facility**" means any structure, group of structures, equipment, or device (other than a vessel) which is used for one or more of the following purposes: exploring for, drilling for, producing, storing, handling, transferring, processing, or transporting oil. This term includes any motor vehicle, rolling stock, or pipeline used for one or more of these purposes;

(10) "**foreign offshore unit**" means a facility which is located, in whole or in part, in the territorial sea or on the continental shelf of a foreign country and which is or was used for one or more of the

following purposes: exploring for, drilling for, producing, storing, handling, transferring, processing, or transporting oil produced from the seabed beneath the foreign country's territorial sea or from the foreign country's continental shelf;

(11) "**Fund**" means the Oil Spill Liability Trust Fund, established by section 9509 of the Internal Revenue Code of 1986 (26 U.S.C. § 9509);

(12) "**gross ton**" has the meaning given that term by the Secretary under part J of title 46, United States Code [46 U.S.C. §§ 14101 *et seq*.];

(13) "**guarantor**" means any person, other than the responsible party, who provides evidence of financial responsibility for a responsible party under this Act;

(14) "**incident**" means any occurrence or series of occurrences having the same origin, involving one or more vessels, facilities, or any combination thereof, resulting in the discharge or substantial threat of discharge of oil;

(15) "**Indian tribe**" means any Indian tribe, band, nation, or other organized group or community, but not including any Alaska Native regional or village corporation, which is recognized as eligible for the special programs and services provided by the United States to Indians because of their status as Indians and has governmental authority over lands belonging to or controlled by the tribe;

(16) "**lessee**" means a person holding a leasehold interest in an oil or gas lease on lands beneath navigable waters (as that term is defined in section 2(a) of the Submerged Lands Act (43 U.S.C. § 1301(a))) or on submerged lands of the Outer Continental Shelf, granted or maintained under applicable State law or the Outer Continental Shelf Lands Act (43 U.S.C. § 1331 *et seq*.);

(17) "**liable**" or "**liability**" shall be construed to be the standard of liability which obtains under section 311 of the Federal Water Pollution Control Act (33 U.S.C. § 1321);

(18) "**mobile offshore drilling unit**" means a vessel (other than a self-elevating lift vessel) capable of use as an offshore facility;

(19) "**National Contingency Plan**" means the National Contingency Plan prepared and published under section 311(d) of the Federal Water Pollution Control Act [33 USCS § 1321(d)], as amended by this Act, or revised under section 105 of the Comprehensive

Environmental Response, Compensation, and Liability Act (42 U.S.C. § 9605);

(20) "**natural resources**" includes land, fish, wildlife, biota, air, water, ground water, drinking water supplies, and other such resources belonging to, managed by, held in trust by, appertaining to, or otherwise controlled by the United States (including the resources of the exclusive economic zone), any State or local government or Indian tribe, or any foreign government;

(21) "**navigable waters**" means the waters of the United States, including the territorial sea;

(22) "**offshore facility**" means any facility of any kind located in, on, or under any of the navigable waters of the United States, and any facility of any kind which is subject to the jurisdiction of the United States and is located in, on, or under any other waters, other than a vessel or a public vessel;

(23) "**oil**" means oil of any kind or in any form, including petroleum, fuel oil, sludge, oil refuse, and oil mixed with wastes other than dredged spoil, but does not include any substance which is specifically listed or designated as a hazardous substance under subparagraphs (A) through (F) of section 101(14) of the Comprehensive Environmental Response, Compensation, and Liability Act (42 U.S.C. § 9601) and which is subject to the provisions of that Act;

(24) "**onshore facility**" means any facility (including, but not limited to, motor vehicles and rolling stock) of any kind located in, on, or under, any land within the United States other than submerged land;

(25) the term "**Outer Continental Shelf facility**" means an offshore facility which is located, in whole or in part, on the Outer Continental Shelf and is or was used for one or more of the following purposes: exploring for, drilling for, producing, storing, handling, transferring, processing, or transporting oil produced from the Outer Continental Shelf;

(26) "**owner or operator**"—

(A) means—

(i) in the case of a vessel, any person owning, operating, or chartering by demise, the vessel;

(ii) in the case of an onshore or offshore facility, any person owning or operating such facility;

(iii) in the case of any abandoned offshore facility, the person who owned or operated such facility immediately prior to such abandonment;

(iv) in the case of any facility, title or control of which was conveyed due to bankruptcy, foreclosure, tax delinquency, abandonment, or similar means to a unit of State or local government, any person who owned, operated, or otherwise controlled activities at such facility immediately beforehand;

(v) notwithstanding subparagraph (B)(i), and in the same manner and to the same extent, both procedurally and substantively, as any nongovernmental entity, including for purposes of liability under section 1002, any State or local government that has caused or contributed to a discharge or substantial threat of a discharge of oil from a vessel or facility ownership or control of which was acquired involuntarily through—

I. seizure or otherwise in connection with law enforcement activity;
II. bankruptcy;
III. tax delinquency;
IV. abandonment; or
V. other circumstances in which the government involuntarily acquires title by virtue of its function as sovereign;

(vi) notwithstanding subparagraph (B)(ii), a person that is a lender and that holds indicia of ownership primarily to protect a security interest in a vessel or facility if, while the borrower is still in possession of the vessel or facility encumbered by the security interest, the person—

(I) exercises decision making control over the environmental compliance related to the vessel or facility, such that the person has undertaken responsibility for oil handling or disposal practices related to the vessel or facility; or

(II) exercises control at a level comparable to that of a manager of the vessel or facility, such that the person has assumed or manifested responsibility—

(aa) for the overall management of the vessel or facility encompassing day-to-day decision making with respect to environmental compliance; or (bb) over all or substantially all of the operational functions (as distinguished from financial or administrative functions) of the vessel or facility other than the function of environmental compliance; and

(B) does not include—

(i) A unit of state or local government that acquired ownership or control of a vessel or facility involuntarily through—

I. seizure or otherwise in connection with law enforcement activity;
II. bankruptcy;
III. tax delinquency;
IV. abandonment; or
V. other circumstances in which the government involuntarily acquires title by virtue of its function as sovereign;

(ii) a person that is a lender that does not participate in management of a vessel or facility, but holds indicia of ownership primarily to protect the security interest of the person in the vessel or facility; or

(iii) a person that is a lender that did not participate in management of a vessel or facility prior to foreclosure, notwithstanding that the person—

I. forecloses on the vessel or facility; and
II. after foreclosure, sells, re-leases (in the case of a lease finance transaction), or liquidates the vessel or facility, maintains business activities, winds up operations, undertakes a removal action under section 311(c) of the Federal Water Pollution Control Act (33 U.S.C. § 1321(c)) or under the direction of an on-scene coordinator appointed under the National Contingency Plan, with respect to the vessel or facility, or takes any other measure to preserve, protect, or prepare the vessel or facility prior to sale or disposition, if the person seeks to sell, re-lease (in the case of a lease finance transaction), or otherwise divest the person of the vessel or facility at the earliest practicable, commercially reasonable time, on commercially reasonable terms, taking into account market conditions and legal and regulatory requirements;

(27) "**person**" means an individual, corporation, partnership, association, State, municipality, commission, or political subdivision of a State, or any interstate body;

(28) "**permittee**" means a person holding an authorization, license, or permit for geological exploration issued under section 11 of the Outer Continental Shelf Lands Act (43 U.S.C. § 1340) or applicable State law;

(29) "**public vessel**" means a vessel owned or bareboat chartered and operated by the United States, or by a State or political subdivision thereof, or by a foreign nation, except when the vessel is engaged in commerce;

(30) "**remove**" or "**removal**" means containment and removal of oil or a hazardous substance from water and shorelines or the taking of other actions as may be necessary to minimize or mitigate damage to the public health or welfare, including, but not limited to, fish, shellfish, wildlife, and public and private property, shorelines, and beaches;

(31) "**removal costs**" means the costs of removal that are incurred after a discharge of oil has occurred or, in any case in which there is a substantial threat of a discharge of oil, the costs to prevent, minimize, or mitigate oil pollution from such an incident;

(32) "**responsible party**" means the following:

a) Vessels. In the case of a vessel, any person owning, operating, or demise chartering the vessel.
b) Onshore facilities. In the case of an onshore facility (other than a pipeline), any person owning or operating the facility, except a Federal agency, State, municipality, commission, or political subdivision of a State, or any interstate body, that as the owner transfers possession and right to use the property to another person by lease, assignment, or permit.
c) Offshore facilities. In the case of an offshore facility (other than a pipeline or a deepwater port licensed under the Deepwater Port Act of 1974 (33 U.S.C. § 1501 *et seq.*)), the lessee or permittee of the area in which the facility is located or the holder of a right of use and easement granted under applicable State law or the Outer Continental Shelf Lands Act (43 U.S.C. §§ 1301- 1356) for the area in which the facility is located (if the holder is a different person than the lessee or permittee), except a Federal agency, State, municipality, commission, or political subdivision of a State, or any interstate body, that as owner transfers possession and right to use the property to another person by lease, assignment, or permit.

d) Deepwater ports. In the case of a deepwater port licensed under the Deepwater Port Act of 1974 (33 U.S.C. §§ 1501-1524), the licensee.
e) Pipelines. In the case of a pipeline, any person owning or operating the pipeline.
f) Abandonment. In the case of an abandoned vessel, onshore facility, deepwater port, pipeline, or offshore facility, the persons who would have been responsible parties immediately prior to the abandonment of the vessel or facility.

(33) "**Secretary**" means the Secretary of the department in which the Coast Guard is operating;

(34) "**tank vessel**" means a vessel that is constructed or adapted to carry, or that carries, oil or hazardous material in bulk as cargo or cargo residue, and that—

a) is a vessel of the United States;
b) operates on the navigable waters; or
c) transfers oil or hazardous material in a place subject to the jurisdiction of the United States;

(35) "**territorial seas**" means the belt of the seas measured from the line of ordinary low water along that portion of the coast which is in direct contact with the open sea and the line marking the seaward limit of inland waters, and extending seaward a distance of 3 miles;

(36) "**United States**" and "**State**" mean the several States of the United States, the District of Columbia, the Commonwealth of Puerto Rico, Guam, American Samoa, the United States Virgin Islands, the Commonwealth of the Northern Marianas, and any other territory or possession of the United States;

(37) "**vessel**" means every description of watercraft or other artificial contrivance used, or capable of being used, as a means of transportation on water, other than a public vessel;

(38) "**participate in management**"—

(A) (i) means actually participating in the management or operational affairs of a vessel or facility; and
(ii) does not include merely having the capacity to influence, or the unexercised right to control, vessel or facility operations; and
(B) does not include—

(i) performing an act or failing to act prior to the time at which a security interest is created in a vessel or facility;
(ii) holding a security interest or abandoning or releasing a security interest;
(iii) including in the terms of an extension of credit, or in a contract or security agreement relating to the extension, a covenant, warranty, or other term or condition that relates to environmental compliance;
(iv) monitoring or enforcing the terms and conditions of the extension of credit or security interest;
(v) monitoring or undertaking one or more inspections of the vessel or facility;
(vi) requiring a removal action or other lawful means of addressing a discharge or substantial threat of a discharge of oil in connection with the vessel or facility prior to, during, or on the expiration of the term of the extension of credit;
(vii) providing financial or other advice or counseling in an effort to mitigate, prevent, or cure default or diminution in the value of the vessel or facility;
(viii) restructuring, renegotiating, or otherwise agreeing to alter the terms and conditions of the extension of credit or security interest, exercising forbearance;
(ix) exercising other remedies that may be available under applicable law for the breach of a term or condition of the extension of credit or security agreement; or
(x) conducting a removal action under 311(c) of the Federal Water Pollution Control Act (33 U.S.C. § 1321(c)) or under the direction of an on-scene coordinator appointed under the National Contingency Plan, if such actions do not rise to the level of participating in management under subparagraph (A) of this paragraph and paragraph (26)(A)(vi);

(39) "**extension of credit**" has the meaning provided in section 101(20)(G)(i) of the Comprehensive Environmental Response, Compensation and Liability Act of 1980 (42 U.S.C. § 9601(20)(G)(i));

(40) "**financial or administrative function**" has the meaning provided in section 101(20)(G)(ii) of the Comprehensive Environmental Response, Compensation and Liability Act of 1980 (42 U.S.C. § 9601(20)(G)(ii));

(41) "**foreclosure**" and "**foreclose**" each has the meaning provided in section 101(20)(G)(iii) of the Comprehensive Environmental Response, Compensation and Liability Act of 1980 (42 U.S.C. § 9601(20)(G)(iii));

(42) "**lender**" has the meaning provided in section 101(20)(G)(iv) of the Comprehensive Environmental Response, Compensation and Liability Act of 1980 (42 U.S.C. § 9601(20)(G)(iv));

(43) "**operational function**" has the meaning provided in section 101(20)(G)(v) of the Comprehensive Environmental Response, Compensation and Liability Act of 1980 (42 U.S.C. § 9601(20)(G)(v)); and

(44) "**security interest**" has the meaning provided in section 101(20)(G)(vi) of the Comprehensive Environmental Response, Compensation and Liability Act of 1980 (42 U.S.C. § 9601(20)(G)(vi)).

APPENDIX B. OIL POLLUTION ACT OF 1990: LIMITATIONS ON LIABILITY

The Oil Pollution Act of 1990 establishes limitations on the liability of responsible parties for certain damages caused by a discharge of oil into the navigable waters of the United States. For the reader's convenience, this appendix provides the liability limitation language as it appears in 33 U.S.C. § 2704.

Limits on liability

(a) General rule. Except as otherwise provided in this section, the total of the liability of a responsible party under section 1002 [33 USCS § 2702] and any removal costs incurred by, or on behalf of, the responsible party, with respect to each incident shall not exceed—

(1) for a tank vessel, the greater of—

A. with respect to a single-hull vessel, including a single-hull vessel fitted with double sides only or a double bottom only, $3,000 per gross ton;
B. with respect to a vessel other than a vessel referred to in subparagraph (A), $1,900 per gross ton; or
C. (i) with respect to a vessel greater than 3,000 gross tons that is—

(I) a vessel described in subparagraph (A), \$22,000,000; or

(II) a vessel described in subparagraph (B), \$16,000,000; or

(ii) with respect to a vessel of 3,000 gross tons or less that is—

(I) a vessel described in subparagraph (A), \$6,000,000; or

(II) a vessel described in subparagraph (B), \$4,000,000;

(2) for any other vessel, \$950 per gross ton or \$800,000, whichever is greater;

(3) for an offshore facility except a deepwater port, the total of all removal costs plus \$75,000,000; and

(4) for any onshore facility and a deepwater port, \$350,000,000.

(b) Division of liability for mobile offshore drilling units.

(1) Treated first as tank vessel. For purposes of determining the responsible party and applying this Act and except as provided in paragraph (2), a mobile offshore drilling unit which is being used as an offshore facility is deemed to be a tank vessel with respect to the discharge, or the substantial threat of a discharge, of oil on or above the surface of the water.

(2) Treated as facility for excess liability. To the extent that removal costs and damages from any incident described in paragraph (1) exceed the amount for which a responsible party is liable (as that amount may be limited under subsection (a)(1)), the mobile offshore drilling unit is deemed to be an offshore facility. For purposes of applying subsection (a)(3), the amount specified in that subsection shall be reduced by the amount for which the responsible party is liable under paragraph (1).

(c) Exceptions.

(1) Acts of responsible party. Subsection (a) does not apply if the incident was proximately caused by—

A. gross negligence or willful misconduct of, or
B. the violation of an applicable Federal safety, construction, or operating regulation by, the responsible party, an agent or employee of the responsible party, or a person acting pursuant to a contractual relationship with the responsible party (except where the sole contractual arrangement arises in connection with carriage by a common carrier by rail).

(2) Failure or refusal of responsible party. Subsection (a) does not apply if the responsible party fails or refuses—

A. to report the incident as required by law and the responsible party knows or has reason to know of the incident;

B. to provide all reasonable cooperation and assistance requested by a responsible official in connection with removal activities; or
C. without sufficient cause, to comply with an order issued under subsection (c) or (e) of section 311 of the Federal Water Pollution Control Act (33 U.S.C. § 1321), as amended by this Act, or the Intervention on the High Seas Act (33 U.S.C. § 1471 *et seq.*).

(3) OCS facility or vessel. Notwithstanding the limitations established under subsection (a) and the defenses of section 1003 [33 U.S.C. § 2703], all removal costs incurred by the United States Government or any State or local official or agency in connection with a discharge or substantial threat of a discharge of oil from any Outer Continental Shelf facility or a vessel carrying oil as cargo from such a facility shall be borne by the owner or operator of such facility or vessel.

(4) Certain tank vessels. Subsection (a)(1) shall not apply to—

a tank vessel on which the only oil carried as cargo is an animal fat or vegetable oil, as those terms are used in section 2 of the Edible Oil Regulatory Reform Act [33 U.S.C. § 2720]; and

a tank vessel that is designated in its certificate of inspection as an oil spill response vessel (as that term is defined in section 2101 of title 46, United States Code) and that is used solely for removal.

(d) Adjusting limits of liability.

(1) Onshore facilities. Subject to paragraph (2), the President may establish by regulation, with respect to any class or category of onshore facility, a limit of liability under this section of less than $350,000,000, but not less than $8,000,000, taking into account size, storage capacity, oil throughput, proximity to sensitive areas, type of oil handled, history of discharges, and other factors relevant to risks posed by the class or category of facility.

(2) Deepwater ports and associated vessels.

A. Study. The Secretary shall conduct a study of the relative operational and environmental risks posed by the transportation of oil by vessel to deepwater ports (as defined in section 3 of the Deepwater Port Act of 1974 (33 U.S.C. § 1502)) versus the transportation of oil by vessel to other ports. The study shall include a review and analysis of offshore lightering practices used in connection with that transportation, an analysis of the volume of oil transported by vessel using those practices, and an analysis of the frequency and volume of oil discharges which occur in connection with the use of those practices.

B. Report. Not later than 1 year after the date of the enactment of this Act [enacted Aug. 18, 1990], the Secretary shall submit to the Congress a report on the results of the study conducted under subparagraph (A).
C. Rulemaking proceeding. If the Secretary determines, based on the results of the study conducted under [this] subparagraph (A), that the use of deepwater ports in connection with the transportation of oil by vessel results in a lower operational or environmental risk than the use of other ports, the Secretary shall initiate, not later than the 180th day following the date of submission of the report to the Congress under subparagraph (B), a rulemaking proceeding to lower the limits of liability under this section for deepwater ports as the Secretary determines appropriate. The Secretary may establish a limit of liability of less than $350,000,000, but not less than $50,000,000, in accordance with paragraph (1).

(3) Periodic reports. The President shall, within 6 months after the date of the enactment of this Act [enacted Aug. 18, 1990], and from time to time thereafter, report to the Congress on the desirability of adjusting the limits of liability specified in subsection (a).

(4) Adjustment to reflect Consumer Price Index. The President, by regulations issued not later than 3 years after the date of enactment of the Delaware River Protection Act of 2006 [enacted July 11, 2006] and not less than every 3 years thereafter, shall adjust the limits on liability specified in subsection (a) to reflect significant increases in the Consumer Price Index.

End Notes

[1] 33 U.S.C. § 2701 *et seq*. Under OPA, the term "navigable waters" means "the waters of the United States, including the territorial sea." 33 U.S.C. § 2701(21).

[2] A recently introduced bill in the Senate, S. 3305, would raise OPA's limitation on liability for offshore facilities from $75 million to $10 billion, if enacted. The House companion bill to S. 3305, H.R. 5214, would amend OPA's limitation on liability for offshore facilities in the same manner as S. 3305, if enacted.

[3] 33 U.S.C. § 1321.

[4] 33 U.S.C. § 2702(b)(1)(B).

[5] 33 U.S.C. § 2702(b)(2)(A)-(F).

[6] The Oil Spill Liability Trust Fund is a federally administered trust fund that is financed by a per-barrel tax on petroleum products produced for consumption within the United States. 26 U.S.C. §§ 4611.

[7] Under OPA, the term "responsible party" refers to the owner or operator of a vessel or facility from which oil is discharged. 33 U.S.C § 2701(32). *See* Appendix A for OPA's definition of the term "responsible party."

[8] Under OPA, the terms "liable" and "liability" are "construed to be the standard of liability which obtains under section 311 of the [Clean Water Act]." Courts have interpreted section 311 of the Clean Water Act as imposing strict liability on parties responsible for the discharge of oil or hazardous substances into the waters of the United States. *See* United States v. New York, 481 F. Supp. 4 (D.N.Y. 1979).

[9] Unlike removal costs, which are uncapped, the responsible party's liability for damages under OPA is limited based on the type of vessel or facility involved, and the amount of oil discharged. 33 U.S.C. §§ 2704(a) and (b).

[10] 33 U.S.C. §§ 2718(a) and (c).

[11] Under OPA, the term "removal costs" means "the costs of removal that are incurred after a discharge of oil has occurred or, in any case in which there is a substantial threat of a discharge of oil, the costs to prevent, minimize, or mitigate oil pollution from such an incident." 33 U.S.C. § 2701(31).

[12] 33 U.S.C. § 2702(a) (emphasis supplied).

[13] 33 U.S.C. §§ 2702(b)(1)(A) and (B). The National Contingency Plan is authorized by Clean Water Act § 311(d). 33 U.S.C. § 1321(d). The implementing regulations promulgated by the Environmental Protection Agency are set forth at 40 CFR § 300.1 *et seq*.

[14] 33 U.S.C. § 2713(a). Under OPA, the term "claim" means "a request, made in writing for a sum certain, for compensation for damages or removal costs resulting from an [oil spill] incident." 33 U.S.C. § 2701(3).

[15] The Oil Pollution Act of 1990 provides federal district courts with "exclusive original jurisdiction over all controversies arising under [OPA]." 33 U.S.C. § 2717(b). State courts "of competent jurisdiction" may consider claims for damages or removal costs arising under OPA or applicable state law. 33 U.S.C. § 2717(c).

[16] 33 U.S.C. § 2713(c). Claims for removal costs must be presented within six (6) years after the date of completion of all removal activities related to the oil spill incident. 33 U.S.C. § 2712(h)(1). A guide for filing claims against the Oil Spill Liability Trust Fund can be found on the U.S. Coast Guard's National Pollution Fund Center website at http://www.uscg.mil/npfc/Claims/default.asp (last visited May 24, 2010).

[17] 33 U.S.C. § 2713(b).

[18] 33 U.S.C. § 2713(b)(1)(A); 33 CFR § 136.103(b)(1).

[19] 33 U.S.C. § 2708 (responsible party entitled to recover against Fund). *See* 33 U.S.C. § 2713(b)(1)(B); 33 CFR § 136.103(b)(2).

[20] 33 U.S.C. § 2713(b)(1)(C); 33 CFR § 136.103(b)(3).

[21] 33 U.S.C. § 2713(b)(1)(D); 33 CFR § 136.103(b)(4). The term "United States claimant" is not specifically defined by OPA or its implementing regulations.

[22] 33 U.S.C. § 2713(e); 33 CFR §§ 136.203(a), (b), and (c). The Federal On-Scene Coordinator of the Deepwater Horizon oil spill in the Gulf of Mexico is U.S. Coast Guard 8th District Deputy Commander, Rear Admiral James A. Watson.

[23] 33 U.S.C. § 2713(e); 33 CFR § 136.205.

[24] *Id.*

[25] Costs incurred for removal activities that are not coordinated with or directed by the Federal On-Scene Coordinator, however, may be presented by any claimant to the responsible party. 33 U.S.C. § 2713(a).

[26] Under OPA, the term "damages" means "damages specified in [33 U.S.C. § 2702(b)], *and includes the costs of assessing these damages*." 33 U.S.C. § 2701(5) (emphasis supplied). The standards and procedures for conducting natural resource damage assessments are set forth in regulations promulgated by the National Oceanic and Atmospheric Administration pursuant to OPA. 33 U.S.C. § 2706(e); 15 C.F.R. §§ 990.10 through 990.66.

[27] 33 U.S.C. § 2702(a) (emphasis supplied).

[28] 33 U.S.C. § 2713(a). Under OPA, the term "claim" means "a request, made in writing for a sum certain, for compensation for damages or removal costs resulting from an [oil spill] incident." 33 U.S.C. § 2701(3).

[29] 33 U.S.C. § 2713(c). Claims for the recovery of damages for an oil spill incident must be presented within three years of the oil spill incident. 33 U.S.C. § 2712(h)(2). A guide for filing claims against the Oil Spill Liability Trust Fund can be found on the U.S. Coast Guard's National Pollution Fund Center website at http://www.uscg.mil/npfc/Claims/default.asp (last visited May 24, 2010).

[30] *See* 33 U.S.C. § 2704(c) (list of exceptions to liability limitations).

[31] *Id.*

[32] *See* South Port Marine, LLC v. Gulf Oil, LP, 234 F.3d 58 (1st Cir. 2000) (OPA displaces maritime-law punitive damages); Clausen v. M/V New Carissa, 171 F. Supp. 2d 1127 (D. Or. 2003) (OPA provides exclusive federal remedy for property damage claims resulting from oil spill, and thus precludes award of punitive damages for any claim for which the act could provide relief).

[33] 33 U.S.C. § 2704(a).

[34] *Id.* For definitions of the types of facilities covered by OPA, *see* Appendix A.

[35] *See* http://www.deepwaterinvestigation.com/go/doc/3043/558647/ (U.S. Coast Guard and Minerals Management Service joint press release referring to Deepwater Horizon oil rig as a mobile offshore drilling unit) (last visited on May 24, 2010).

[36] 33 U.S.C. § 2704(b)(1). The Federal On-Scene Coordinator designates the source of the discharge for the purposes of determining liability under OPA. 33 U.S.C § 2714(a); 33 C.F.R. § 136.305. The Federal On-Scene Coordinator of the ongoing oil spill in the Gulf of Mexico is U.S. Coast Guard 8th District Deputy Commander, Rear Admiral James A. Watson.

[37] 33 U.S.C. § 2704(b)(2).

[38] For the purposes of determining the liability limitation for a specific vessel or facility, OPA uses a formula that is based on the weight of the vessel or facility owned or operated by the responsible party. *See* Appendix B for OPA limitations on liability.

[39] 33 U.S.C. § 2704(b)(2).

[40] 33 U.S.C. §§ 2704(c)(1)(A) and (B).

[41] 33 U.S.C. §§ 2704(c)(2)(A), (B), and (C).

[42] 33 U.S.C. § 2702(b)(2)(A).

[43] 33 U.S.C. § 2701(20).

[44] 33 U.S.C. § 2702(b)(2)(A).

[45] 33 U.S.C. § 2713(e); 33 CFR Part 136, subparts B and C.

[46] 33 U.S.C. § 2702(b)(2)(B).

[47] 33 U.S.C. § 2702(b)(2)(B).

[48] 33 U.S.C. § 2713(e); 33 CFR §§ 136.215(a)(1)-(4).

[49] 33 U.S.C. § 2713(e); 33 CFR §§ 136.215(b)(1)-(3).

[50] 33 U.S.C. § 2702(b)(2)(C).

[51] *Id.*

[52] 33 U.S.C. § 2702(b)(2)(D).
[53] *Id.*
[54] 33 U.S.C. § 2713(e); 33 CFR §§ 136.227(a)-(d).
[55] 33 U.S.C. § 2713(e); 33 CFR § 136.229.
[56] 33 U.S.C. § 2702(b)(2)(E).
[57] 33 U.S.C. § 2701(20).
[58] 33 U.S.C. § 2713(e); 33 CFR §§ 136.233(a)-(d).
[59] 33 U.S.C. § 2713(e); 33 CFR § 136.235.
[60] 33 U.S.C. § 2702(b)(2)(F).
[61] 33 U.S.C. § 2702(b)(2)(F).
[62] 33 U.S.C. § 2713(e); 33 CFR §§ 136.239(a)-(d).
[63] 33 U.S.C. § 2713(e); 33 CFR § 136.241.
[64] 33 U.S.C. § 2703(a). For the purposes of determining whether the "act of God" complete defense is available to a responsible party, the term "act of God" means an "unanticipated grave natural disaster or other natural phenomenon of an exceptional, inevitable, and irresistible character the effects of which could not have been prevented or avoided by the exercise of due care or foresight." 33 U.S.C. § 2701(1). In the event that the upcoming hurricane season exacerbates the Deepwater Horizon oil spill, the question arises as to whether a hurricane in the Gulf of Mexico could be considered an "unanticipated" natural phenomenon. Analysis and consideration of this question, however, are beyond the scope of this CRS report.
[65] 33 U.S.C. §§ 2703(a)(3)(A) and (B).
[66] 33 U.S.C. § 2703(c).
[67] 33 U.S.C. § 2703(d)(5).
[68] 33 U.S.C. § 2712. The standards and procedural requirements for claims filed against the Fund are set forth in the Coast Guard's OPA regulations. *See* 33 C.F.R. §§ 136.1 through 136.241.
[69] 26 U.S.C. §§ 4611(a)(1) and (2). The Oil Spill Liability Trust fund is also financed by a per-barrel tax on domestic crude oil "used in or exported from the United States." 26 U.S.C. § 4611(b)(1)(A).
[70] 26 U.S.C. § 9509(c)(2)(A).
[71] 26 U.S.C. § 9509(d)(3)(B). Senate bill S. 3306 would reinstate the Fund's authority to receive advances from the Treasury, if enacted. The House companion bill to S. 3306, H.R. 5214, would amend the Fund's authority to borrow in the same manner as S. 3306, if enacted.
[72] 33 U.S.C. § 2715(c).

INDEX

A

abatement, ix, 29, 30, 33, 35, 42
access, 80, 88, 92
achieve higher and higher, ix, 29, 33
acid, 68
adjoining shorelines, xii, xiii, 135, 136, 137, 138
adjustment, 116
administering, xi, 9, 16, 47, 48, 61, 88
age, 114
agencies, viii, x, xi, xii, 2, 4, 7, 8, 9, 11, 12, 14, 18, 19, 20, 22, 26, 46, 47, 48, 49, 52, 58, 61, 76, 83, 84, 86, 87, 90, 94, 96, 102, 106, 107, 111, 112, 120, 121, 122, 123, 124, 132
air pollutants, 129
Alaska, 4, 14, 40, 100, 103, 127, 132, 146
alters, 26
ambient air, 6, 7, 63
American Samoa, 151
appropriations, xi, 18, 31, 37, 43, 46, 48, 50, 52, 56, 62, 72, 75, 77, 82, 83, 86, 88, 89, 91, 93, 94, 95, 127
Area Committees (ACs), 14, 15, 18, 121
asbestos, 55
assessment, 34, 71, 77, 94, 111, 141
assets, 67
atmospheric deposition, 104, 105, 129
Attorney General, 144

B

background information, xi, 4, 98, 102, 103, 129
bacteria, ix, 30, 33, 37
bankruptcy, 66, 80, 148, 149
base, 64, 117
batteries, 68
benefits, 114
birds, 109
BOEMRE, 132
budget deficit, 36
Bureau of Ocean Energy Management, Regulation, and Enforcement (BOEMRE), 117, 132
businesses, 73, 74, 112, 116

C

cadmium, 68
Caribbean, 14
case law, 64
CDC, 5, 12, 25
CERCLA for specific purposes, x, 45
certificate, 155
certification, 41
CFR, 157, 158, 159
challenges, 12, 49, 57, 104
chemical(s), vii, 1, 3, 18, 32, 34, 35, 38, 73, 74

cities, 33, 36, 37
citizens, 81
civil action, 81, 82
classes, 38
Clean Water Act (CWA), v, vii, ix, xii, 2, 3, 4, 5, 6, 7, 8, 14, 15, 17, 22, 23, 24, 27, 29, 30, 31, 32, 34, 36, 40, 42, 54, 60, 64, 102, 113, 118, 121, 126, 127, 129, 136, 137, 157
cleaning, 35, 47, 88, 95
Coast Guard, viii, xii, 2, 7, 8, 9, 11, 12, 15, 16, 17, 18, 22, 24, 25, 26, 52, 76, 102, 106, 107, 113, 115, 116, 117, 121, 123, 124, 125, 129, 131, 151, 157, 158, 159
coastal region, 121
combustion, 129
commerce, 120, 150
commercial, 68, 104, 105, 111
communication, 17
community(s), x, 20, 30, 37, 55, 61, 93, 110, 146
compensation, xii, 102, 125, 126, 138, 141, 142, 143, 145, 157, 158
complexity, 4, 10, 16, 21, 109
compliance, 33, 37, 38, 60, 68, 69, 76, 81, 93, 98, 148, 149, 152
Comprehensive Environmental Response, Compensation, and Liability Act (CERCLA), v, vii, x, 2, 4, 5, 6, 23, 24, 26, 27, 45, 47, 53, 99, 121, 130, 131, 147
Comprehensive Environmental Response, Compensation, and Liability Act of 1980, x, 4, 23, 45, 47, 53
conference, 125, 126
Conference Report, 131, 133
Congressional Budget Office, 39
congressional interest, xii, 135
consensus, 85
consent, 32, 123
Consolidated Appropriations Act, 67, 68
constituents, 92
Constitution, 99
construction, ix, x, 29, 30, 32, 33, 34, 36, 38, 41, 58, 78, 114, 119, 125, 131, 139, 154
consumer price index, 115, 131, 156
consumption, 62, 104, 105, 144, 156
containers, 68
contaminant, 9, 18, 19, 25, 49, 52, 53, 67, 96
contaminated food, 62
contaminated sites, x, 4, 6, 45, 46, 47, 48, 49, 57, 59, 67, 72, 74, 86, 95
contamination, x, 25, 46, 47, 49, 50, 51, 52, 55, 56, 57, 59, 62, 64, 65, 66, 70, 71, 72, 73, 74, 81, 83, 84, 87, 88, 89, 90, 91, 92, 93, 94
Continental, 146, 147, 155
contingency, 23, 122
controversial, 38, 74, 108
controversies, 157
convention, 125, 133
conversations, 129
cooling, 112
cooperation, 40, 144, 155
cooperative agreements, 82
coordination, 4, 7, 9, 16, 17, 21, 22, 42
coral reefs, 109, 110
cost, xi, 46, 48, 56, 57, 58, 59, 64, 72, 75, 76, 77, 78, 80, 83, 91, 110, 111, 114, 130, 132, 138, 140, 143, 145
counseling, 152
covering, 124
CPI, 115
CRS report, 159
crude oil, 24, 53, 103, 110, 120, 128, 133, 144, 159
culture, 110
cure, 152
CWA, 5, 33, 34, 35, 36, 37, 38, 39, 113, 114, 118, 127, 130, 131, 132, 133

D

damages, x, xii, xiii, 4, 5, 19, 46, 62, 63, 64, 67, 76, 77, 110, 111, 112, 115, 116, 117, 122, 127, 129, 135, 136, 137, 138, 139, 140, 141, 142, 143, 144, 145, 153, 154, 157, 158
danger, 6, 39, 52, 125

database, 25, 129
days to years, xi, 101, 108
deduction, 95
Deepwater Horizon, xi, xii, 4, 12, 16, 17, 21, 24, 25, 26, 101, 102, 103, 106, 108, 116, 122, 126, 130, 131, 135, 139, 157, 158, 159
delegates, 8, 133
delinquency, 66, 148, 149
demonstrations, 40
Department of Agriculture, 11
Department of Commerce, 11
Department of Defense, 8, 11, 48, 50, 57, 88, 91, 92, 93, 99
Department of Energy, 8, 11, 48, 57, 93, 99
Department of Health and Human Services, 5, 11, 12
Department of Homeland Security, 5, 16, 21, 25, 26, 27, 133
Department of Justice, 11, 12, 76, 77, 130
Department of Labor, 11
Department of the Interior, 11, 27, 130, 132
Department of Transportation, 11, 25, 119, 123
deposition, 105, 129
deposits, 120
depth, 24, 57
derivatives, 73
destruction, 63, 64, 65, 77, 115, 140, 141, 142
DHS, 5
direct action, 39
directives, 27
disaster, 19, 20, 103, 145, 159
discharges, vii, viii, ix, xii, 1, 2, 3, 4, 6, 7, 8, 9, 12, 15, 19, 20, 21, 22, 25, 26, 29, 30, 33, 34, 35, 37, 42, 104, 105, 109, 112, 118, 124, 129, 135, 136, 139, 155
disposition, 149
dissolved oxygen, 33
district courts, 157
District of Columbia, 25, 36, 82, 151
DOC, 11
DOI, 11, 132
DOJ, 11
DOL, 11
DOT, 11, 119
drainage, 35
drinking water, 55, 60, 111, 140, 142, 147
dumping, 129, 145

E

earnings, 142
economic damage(s), 19
economic losses, 19, 110, 115, 140, 141
ecosystem, 104, 108, 109
effluent, ix, 30, 33, 35, 37, 38, 39, 41
emergency, vii, viii, x, 1, 2, 3, 6, 8, 13, 19, 20, 25, 26, 27, 46, 55, 56, 62, 86, 93, 124, 131
Emergency Assistance, 19
Emergency Economic Stabilization Act, 117
emergency planning, 13
Emergency Planning and Community Right-to-Know Act (EPCRA), 13
emergency response, vii, viii, 1, 2, 3, 13, 26, 27
emission, 145
employment, 142
energy, 69, 102, 126
Energy Policy Act of 2005, 117
energy recovery, 69
enforcement, ix, x, 6, 23, 30, 32, 33, 35, 38, 39, 41, 45, 46, 49, 50, 52, 54, 73, 76, 77, 78, 80, 81, 82, 83, 84, 91, 121, 148, 149
engineering, 60
England, 3, 24, 113, 127
environment(s), vii, ix, x, 1, 3, 4, 6, 7, 12, 15, 23, 24, 25, 30, 33, 46, 47, 48, 52, 55, 59, 60, 62, 63, 72, 76, 77, 82, 84, 85, 88, 90, 91, 92, 104, 110, 112, 125, 130
environmental contamination, 25, 47, 51, 52, 74, 95, 99
environmental impact, 103, 104, 111
environmental protection, 32
Environmental Protection Agency, viii, x, xii, 2, 5, 11, 22, 24, 27, 39, 46, 47, 96, 97, 98, 102, 123, 157
EPCRA, 23, 25

equipment, 3, 12, 17, 23, 114, 118, 119, 122, 124, 130, 145
Europe, 130
evacuation, 103
evidence, 61, 65, 86, 109, 117, 143, 146
evil, 103
examinations, 123
exclusion, 74, 96
execution, 4
executive branch, 16, 27
Executive Order(s), viii, 2, 3, 4, 7, 8, 15, 22, 23, 24, 25, 27, 52, 76, 85, 96, 116, 122, 130, 132, 133
exercise, 69, 145, 159
expenditures, 19, 26
expertise, viii, 2, 12
expired industry taxes, xi, 46, 48
exposure, 51, 52, 54, 56, 57, 59, 62, 66, 71, 82, 87, 90, 92, 108, 109
extraction, 104, 105, 108, 123, 127, 129

F

fails to settle such claims, xiii, 136
fairness, 50, 65, 73, 74, 81
farm land, ix, 30
fat, 155
federal agency, viii, 2, 17, 20, 52, 57, 61, 83, 84, 85, 88, 90, 111, 114
federal aid, 36
federal assistance, x, 16, 20, 32, 45, 49, 51
federal attention, vii, 1, 3, 9
federal authorities, 127
Federal Emergency Management Agency, 11, 27
federal facility(s), viii, x, xi, 8, 12, 15, 18, 23, 45, 46, 47, 48, 49, 50, 52, 56, 57, 58, 61, 81, 82, 83, 84, 86, 87, 88, 91, 93, 99
federal funds, 15
federal government, ix, x, xi, 4, 9, 13, 18, 19, 20, 30, 35, 39, 45, 47, 48, 58, 70, 72, 75, 76, 78, 80, 81, 83, 88, 101, 113, 120, 121, 125, 127, 136
federal law, x, 38, 39, 46, 54, 61, 67, 118, 119, 124
Federal Register, 22, 24, 25, 27, 96, 130, 131, 132
federal regulations, 66, 117
feedstocks, 18, 73, 74
FEMA, 11, 26
financial, ix, 29, 33, 35, 49, 51, 58, 63, 64, 66, 67, 70, 72, 74, 76, 77, 79, 80, 88, 93, 94, 95, 116, 117, 118, 146, 149, 152
financial condition, 79
fires, 3
first responders, vii, 1, 3, 9, 13
fish, 33, 109, 111, 140, 142, 147, 150
fishing, 107, 111, 112, 129
Flow Rate Technical Group, 103, 129, 132
force, 3, 124, 133
foreclosure, 66, 67, 148, 149, 153
formula, 36, 158
framework, vii, viii, xi, xii, 2, 3, 4, 9, 13, 27, 102, 103, 112, 121, 135, 136
France, 24
funding, viii, xi, 2, 3, 4, 5, 18, 33, 36, 46, 50, 52, 57, 61, 75, 82, 83, 86, 90, 94, 100, 111, 112, 113, 116, 117
funds, x, xi, 8, 17, 18, 19, 30, 31, 36, 37, 43, 46, 48, 57, 72, 75, 78, 83, 88, 94, 99, 100

G

GAO, 40, 99
General Accounting Office, 99
General Services Administration, 11
generated or transported, xi, 46, 69
God, 65, 115, 143, 145, 159
governance, vii, xi, xii, 101, 102, 107, 126
government revenues, 141
governments, 13, 86, 111, 121, 140
governor, 13, 90, 111, 113, 122
grants, x, 30, 35, 36, 37, 40, 41, 51, 61, 82, 93, 94, 95, 100
Great Britain, 125
groundwater, 6, 35, 55, 56, 58, 62, 63, 83
GSA, 11
guidance, 38, 49, 79, 97, 124
guidelines, 31, 38, 41
guiding principles, 20

Gulf Coast, 108, 126, 132
Gulf of Mexico, xi, xii, 4, 12, 16, 27, 101, 103, 104, 116, 119, 130, 132, 135, 139, 157, 158, 159

H

habitats, 109, 110, 111
hair, 11
harbors, 7, 35, 52, 76
Hawaii, 14
hazardous materials, 20
Hazardous Substance Superfund Trust Fund, xi, 18, 26, 46, 48, 72
hazardous substances, vii, viii, ix, x, 1, 2, 3, 4, 6, 7, 8, 9, 15, 17, 19, 20, 21, 22, 23, 27, 30, 35, 46, 50, 53, 54, 55, 59, 62, 63, 64, 66, 69, 71, 74, 79, 82, 91, 115, 118, 121, 157
hazardous waste(s), 25, 87, 92, 99
hazards, 4, 9, 10, 20, 25, 62, 92, 142
health, 12, 42, 56, 59, 62, 77, 82, 95, 142, 150
Health and Human Services, 82
health effects, 62, 77
health risks, 62, 82
heavy metals, 38
HHS, 5, 11, 12
history, 24, 64, 108, 129, 132, 155
host, 60
hotels, 116
House, 130, 131, 132, 134, 156, 159
human, 24, 25, 47, 48, 55, 59, 62, 72, 77, 84, 85, 88, 90, 94, 104, 112
human behavior, 104
human exposure, 59, 62
human health, 24, 25, 47, 48, 59, 62, 72, 77, 84, 85, 88, 90, 94, 112
hydrocarbons, 104, 129

I

identification, 85
immunity, 67
IMO, 124, 133
impairments, 38
implemented, ix, 14, 29, 30, 107, 119
imprisonment, 38
improvements, ix, 30
incidents, vii, viii, 1, 2, 4, 5, 6, 8, 9, 12, 13, 15, 16, 19, 20, 21, 25, 26, 106, 107, 133
income, 18, 73, 74, 75, 95, 96, 142
income tax, 74, 75, 95
Indians, 146
individual action, 87
individuals, 15, 39, 52, 56, 62, 110, 112
industry(s), ix, xi, 18, 29, 30, 33, 36, 37, 38, 46, 48, 50, 75, 103, 112, 114, 117, 125, 126, 128
infrastructure, 112
initiation, 78
injury(s), 39, 62, 64, 65, 67, 77, 115, 116, 140, 141, 142
inland zone, viii, xii, 2, 7, 8, 9, 12, 15, 17, 102, 120, 121, 133
inspections, 152
integrity, 32
international affairs, 120
International Maritime Organization (IMO), 124
international standards, 114, 128
investments, 33
involving a discharge, viii, 2, 5, 15
involving pipelines, xii, 102, 126
iron, 35
issues, xi, xii, 4, 31, 35, 73, 101, 102, 103, 112, 116, 119, 124, 125, 126, 127, 128, 129, 131, 133

J

judicial interpretation, 64
jurisdiction(s), ix, 6, 8, 18, 30, 32, 35, 60, 76, 80, 83, 84, 85, 88, 91, 106, 107, 113, 119, 120, 122, 123, 124, 126, 132, 147, 151, 157

K

kill, 108

L

laboratory studies, 109
Lake Pontchartrain, 40
lakes, 42
larvae, 109
law enforcement, 148, 149
laws, vii, x, xii, 32, 35, 39, 40, 45, 47, 53, 54, 68, 69, 87, 102, 108, 113, 118, 119, 121, 125, 126
lead, viii, 2, 8, 15, 16, 21, 33, 68, 84, 109, 111
lead agency, viii, 2, 8, 15
legal issues, 64
legislation, ix, 4, 29, 30, 32, 50, 90, 95, 108, 114, 115, 117, 118, 119, 123, 125, 127, 128, 132, 133
light, 74, 143
litigation, 12, 78
loans, 37, 93, 94
local authorities, 9, 16, 20
local government, 4, 9, 13, 16, 19, 31, 36, 66, 84, 86, 94, 121, 140, 147, 148, 149
location of the spill, xi, xii, 101, 102, 108, 121
Louisiana, 123, 132
Love Canal, 4, 47

M

magnitude, 4, 10, 16, 21, 129
majority, 38, 39, 56, 99, 104, 129
mammals, 109
man, 104
management, 6, 16, 34, 35, 38, 41, 42, 67, 99, 148, 149, 151, 152
manufacturing, 35, 73, 100
marine environment, 113, 124
materials, x, 20, 27, 33, 34, 37, 41, 46, 51, 54, 66, 68, 69, 110, 133
media, 27, 110
medical, 62
medical care, 62
metals, 38
Mexico, xii, 36, 103, 135, 139, 157, 158, 159
migration, 51, 55, 66, 69, 70, 71, 72, 92
military, 8, 25, 50, 88, 89, 90, 91, 92, 100
Minerals Management Service (MMS), 117, 132, 158
mission(s), 12, 25
Missouri, 47
MMS, 117, 132
Mobile offshore drilling units, 116
modifications, 22, 24
Montana, 120
municipal solid waste, 69, 70

N

National Academy of Sciences, 131
National Contingency Plan, vii, 1, 3, 5, 14, 21, 55, 113, 130, 133, 137, 138, 139, 146, 149, 152, 157
National Defense Authorization Act, 50, 53, 90, 92
National Priorities List, x, 5, 6, 26, 46, 48, 55, 84, 86
National Research Council, 104, 105, 123, 128, 131
National Response Framework, viii, 2, 4, 5, 20, 27
National Response System (NRS), viii, 2, 9, 11, 121
National Response Team (NRT), viii, 2, 5, 8, 11, 121
national security, 93
national standards, ix, 30, 34, 35
natural disaster(s), 20, 145, 159
natural resource damages, 140
natural resources, 6, 63, 64, 65, 77, 109, 111, 112, 113, 115, 116, 120, 122, 131, 140, 141, 142, 147
NCP, vii, viii, 1, 2, 3, 4, 5, 6, 7, 8, 9, 11, 12, 13, 14, 15, 16, 17, 18, 20, 21, 22, 23, 24,

25, 26, 27, 55, 56, 77, 113, 114, 121, 127, 130, 134
negotiating, 85
negotiation, 79, 80, 85
nickel, 68
NOAA, 27, 103, 109, 111, 112, 122, 129, 130, 133
non-governmental entities, 17, 18
nonpoint source, ix, 30, 34
North America, 105, 108, 129
not listed on the NPL, x, 46, 69, 84, 86, 87, 90, 93
NPL, x, 5, 6, 26, 46, 48, 51, 55, 56, 57, 60, 61, 62, 69, 70, 84, 85, 86, 87, 90, 93, 94, 97
NRC, 105, 128, 129
NRDA, 130
NRF, viii, 2, 4, 5, 20, 21, 26
NRF itself is not an operational, viii, 2, 20
NRS, viii, 2, 9, 121
NRT, 5, 10, 12, 13, 121, 122
Nuclear Regulatory Commission, 11
nuclear weapons, 25
nutrients, 34

O

OCS, 132, 155
Office of Management and Budget, 8
officials, vii, viii, 1, 2, 3, 9, 13, 15, 52, 86, 111, 131
oil and chemical spills, vii, 1, 3
oil or release of a hazardous substance, viii, 2, 18, 21
Oil Pollution Act (OPA), v, vii, viii, xii, 2, 4, 5, 6, 23, 25, 26, 54, 107, 111, 112, 113, 118, 121, 130, 131, 133, 135, 136, 137, 144, 153, 157
Oil Pollution Act of 1990 (OPA), v, vii, xii, 2, 4, 6, 23, 25, 26, 54, 107, 112, 113, 118, 130, 131, 133, 135, 136, 137, 144, 153, 157
Oil Spill Liability Trust Fund (OSLTF), viii, xiii, 2, 18, 26, 74, 98, 111, 116, 136, 137, 138, 140, 141, 142, 144, 146, 156, 157, 158
OMB, 8
omission, 65, 115, 143
On-Scene Coordinator (OSC), viii, 2, 5, 15, 25, 113, 121, 138, 157, 158
OPA limits certain liabilities, xii, 135, 136
operations, 18, 79, 100, 108, 112, 119, 122, 149, 151
opportunities, 61, 86
organic chemicals, 38
OSC, 5, 10, 15, 17, 19, 120, 121, 122
OSLTF, 117
Outer Continental Shelf, 119, 127, 146, 147, 150, 155
Outer Continental Shelf Lands Act, 119, 127, 146, 150
overlap, 129
oversight, 4, 39, 52, 122
ownership, 50, 51, 66, 67, 70, 71, 72, 87, 88, 89, 90, 93, 140, 144, 148, 149
oxygen, 33, 37

P

P.L. 107-118, 51, 53, 69, 70, 71, 79, 94
P.L. 99-499, x, 23, 24, 45, 50, 53
Pacific, 14
participants, 81
pathways, 62
penalties, xi, xii, 35, 38, 46, 75, 78, 82, 102, 126, 132
permit, ix, 30, 36, 37, 38, 39, 81, 92, 129, 150
persons injured, xii, 135
pesticide, 54, 66
petroleum, x, 4, 35, 46, 50, 53, 73, 74, 94, 96, 103, 104, 110, 128, 129, 144, 147, 156
pH, 33
physical structure, 109
pipeline, 102, 119, 120, 123, 127, 132, 145, 150, 151
plants, ix, 30, 33, 36, 111, 112
platform, 103, 105

policy, xii, 40, 79, 87, 88, 95, 102, 105, 126
policymakers, 104
pollutants, vii, ix, 1, 3, 4, 6, 7, 20, 21, 23, 30, 32, 33, 34, 35, 37, 38, 40, 47, 53, 55, 109
polluters, 48, 73
pollution, ix, 12, 23, 29, 30, 31, 32, 33, 34, 35, 36, 38, 40, 42, 43, 47, 107, 112, 115, 119, 124, 125, 128, 131, 133, 150, 157
polychlorinated biphenyls (PCBs), 68
population, 36
potential sources, xii, 60, 102, 122
precipitation, 129
preparation, 3, 34, 105
preparedness, xii, 102, 113, 118, 121, 122, 123, 126
President, 5, 6, 7, 8, 12, 14, 15, 16, 20, 21, 23, 24, 27, 52, 55, 59, 67, 76, 82, 85, 93, 111, 113, 115, 117, 119, 126, 137, 144, 155, 156
presidential declaration, 19
prevention, xi, xii, 35, 102, 103, 105, 113, 118, 119, 121, 122, 124, 126
primary function, 14, 121
Prince William Sound, 103
prioritizing cleanup, x, 46
private entities, viii, 2, 4, 10, 122, 141
private party, 115
procurement, 42
profit, 69, 141
project, 36, 37, 40
protection, 35, 37, 38, 42, 47, 60, 70, 78, 114, 142
public health, x, 6, 12, 19, 35, 46, 52, 55, 56, 60, 62, 63, 76, 91, 150
public interest, 32
public sector, 95
public service, 19, 115, 116, 142, 143
Puerto Rico, 151

Q

qualifications, 126
quality standards, 32, 34, 41
query, 97

R

rainfall, 34
real property, 141, 142
recovering removal costs, xiii, 136
recovery, 64, 76, 77, 78, 80, 83, 91, 103, 108, 109, 116, 122, 132, 139, 158
recreation, 34, 42, 111
recreational, 104, 105, 107, 111, 129
recycling, 36, 51, 66, 68
redevelopment, x, 46, 47, 49, 51, 89, 93, 94, 95
Reform, 155
Regional Response Teams (RRTs), 13, 14, 15, 121
Registry, 5, 12, 19, 56, 62
regulations, xii, xiii, 3, 6, 7, 8, 15, 21, 23, 26, 27, 35, 38, 49, 55, 57, 60, 71, 92, 102, 106, 107, 109, 111, 112, 115, 118, 119, 121, 122, 124, 132, 136, 139, 140, 156, 157, 158, 159
regulatory framework, 103
regulatory requirements, ix, 29, 33, 38, 149
relief, x, 45, 49, 51, 158
remedial actions, 25, 56, 57, 58, 59, 60, 61, 72, 84, 85, 86, 88
remediation, 6, 24, 26, 37, 58, 94
removal actions, 56, 57, 58, 61, 78, 86, 93, 137
reproduction, 108
reputation, 112
resolution, 85
resources, vii, viii, x, 1, 2, 3, 4, 9, 10, 15, 16, 17, 18, 19, 20, 23, 26, 39, 46, 62, 73, 95, 98, 111, 112, 113, 116, 120, 122, 140, 141, 142, 147
responsible party, xii, xiii, 17, 20, 48, 57, 76, 77, 78, 79, 80, 82, 115, 117, 120, 122, 135, 136, 137, 138, 139, 140, 141, 142, 143, 144, 146, 150, 153, 154, 157, 158, 159
restoration, xii, 32, 102, 109, 111, 112, 127, 133
RESTORE Act, xii, 102, 126
restrictions, 72, 89

restructuring, 152
revenue, 73, 116, 132, 141, 142
risk(s), 25, 48, 56, 57, 59, 60, 61, 62, 84, 86, 90, 91, 92, 93, 94, 95, 114, 122, 155, 156
RRTs, 122
rubber, 68
rule of law, 137, 138
rules, 127
runoff, ix, 30, 34, 38, 104, 105, 109, 129

S

safety, 13, 25, 91, 119, 124, 131, 139, 142, 154
Samoa, 151
SARA, x, 5, 6, 7, 23, 25, 45, 50
savings, 125
scope, vii, 1, 3, 4, 5, 10, 16, 20, 21, 49, 53, 57, 58, 64, 87, 91, 92, 94, 100, 104, 115, 118, 159
Secretary of Defense, 91
Secretary of Homeland Security, 16, 20, 21, 25
security, 67, 93, 119, 148, 149, 152, 153
security practices, 119
sediments, 34
seed, 93
seizure, 148, 149
Senate, 123, 156, 159
separate operational elements, viii, 2
services, 109, 111, 112, 142, 143, 146
settlements, 75, 79, 81, 83
sewage, ix, 29, 30, 33, 36, 38, 42
shape, 30
shellfish, 33, 150
shoreline(s), vii, 1, 3, 110, 118, 137, 138, 150
shortfall, xi, 46, 48
showing, 33
significant, xi, xii, 33, 56, 61, 101, 102, 103, 104, 108, 126, 135, 156
size of the spill, xi, 101, 108
skilled personnel, 12
sludge, 9, 38, 42, 147
Small Business Liability Relief and Brownfields Revitalization Act, 51, 53, 69, 94
small businesses, 69
solid waste, 69
solution, 57
SPCC, 118, 132
species, 109
specifications, 41
spending, 36
spill prevention control and countermeasure (SPCC), 118
stability, 114
Stafford Act, 19, 20, 26
stakeholders, 4
state authorities, 84
state laws, 125, 128, 133
states primarily, x, 46
statistics, 103
statutes, vii, viii, xii, 2, 3, 4, 5, 7, 54, 60, 102, 112, 115, 125, 126
statutory authority, 3, 5, 22, 51, 92, 94
statutory provisions, 49
steel, 35
storage, 99, 100, 118, 155
stormwater, 34, 37
structure, 23, 55, 74, 109, 117, 127, 145
Submerged Lands Act (SLA), 120, 146
subpoena, 80
subsistence, 115, 141
Superfund, v, vii, viii, x, xi, 2, 5, 6, 18, 19, 23, 24, 26, 27, 45, 46, 47, 48, 49, 50, 51, 52, 53, 56, 57, 58, 60, 67, 68, 72, 73, 74, 75, 77, 81, 83, 88, 91, 93, 94, 95, 96, 97, 98, 100, 115, 121, 131
Superfund Amendments and Reauthorization Act of 1986, x, 5, 6, 23, 45, 50, 53, 67, 73, 81, 83, 91, 96, 100
Superfund program, xi, 6, 46, 47, 48, 50, 52, 72, 73, 75, 93, 94, 95, 97
supervision, 94
support, xii, 12, 14, 19, 20, 31, 35, 47, 75, 82, 93, 95, 102, 113, 117, 122, 127
Supreme Court, 125, 126, 129
surplus, 50, 88, 89, 90

survival, 109

T

tanks, 54, 74
tax deduction, 95
tax incentive, 95, 96
taxes, xi, 18, 46, 48, 50, 72, 73, 74, 141
taxpayers, 74
teams, 11, 14, 21
technical assistance, 31, 61
technology(s), ix, 29, 32, 33, 34, 36, 37, 38, 82, 88, 89
telephone, 27
territorial, 13, 14, 15, 24, 145, 147, 151, 156
territory, 13, 151
terrorist attack, 16
testing, 62
textiles, 68
threats, 12, 22
time frame, 85
timing and location, xi, 101, 108
Title I, 23, 25, 32, 36, 50, 53, 67, 69, 79, 94
Title II, 23, 25, 32, 36, 50, 53, 67, 69, 94
Title V, 32, 37, 50, 51, 53, 68, 74, 96, 131
total costs, 143
total revenue, 73
totally revised, ix, 29, 30
toxic substances, 33, 62
toxic waste, 65
toxicity, 59, 82, 108
trade, 112
training, 40, 50, 82, 89, 91, 92, 122, 126
transactions, 81
transmission, 123
transport, 63, 66, 69, 103
transportation, 102, 104, 105, 118, 123, 128, 132, 151, 155, 156
Treasury, 48, 50, 75, 111, 126, 144, 159
treatment, ix, 29, 30, 32, 33, 35, 36, 37, 41, 52, 56, 59, 63, 96, 99
triggers, 131
Trust Fund, viii, xi, xiii, 2, 18, 19, 26, 46, 47, 48, 49, 50, 60, 72, 73, 74, 75, 77, 83, 98, 102, 116, 117, 127, 136, 137, 138, 140, 141, 142, 144, 146, 156, 157, 158
two categories, xii, 69, 102, 121

U

U.S. Army Corps of Engineers, 38
U.S. Department of Agriculture, 11
U.S. Geological Survey, 39
U.S. Treasury, 18, 48, 73, 74, 83
uniform, 40
United Nations, 124
United States coastal waters, vii
unlimited liability, 107, 125, 128
urban, ix, 30, 34, 105, 109, 129
urban areas, ix, 30, 34
USDA, 11

V

variables, 109, 110
vegetable oil, 155
vegetation, 89
vehicles, 131, 147
vessel or facility involved, xii, 135, 136, 157
vessels, xii, 8, 12, 15, 17, 23, 102, 104, 105, 107, 114, 115, 116, 117, 119, 122, 124, 125, 128, 129, 132, 133, 139, 146, 155

W

waiver, 60, 80
war, 65, 115, 143
Washington, 26, 126
waste, x, 33, 41, 46, 47, 49, 51, 63, 64, 66, 69, 87, 92
waste disposal, x, 46, 47, 92
waste disposal sites, x, 46, 47
waste management, 87
waste treatment, 41
wastes discharged, ix, 30, 34
wastewater, x, 30, 32, 33, 35, 36

water, ix, 6, 9, 29, 30, 31, 32, 33, 34, 35, 36, 37, 38, 40, 41, 42, 55, 56, 58, 60, 62, 63, 83, 104, 109, 110, 112, 122, 124, 129, 131, 133, 140, 142, 147, 150, 151, 154
water quality, ix, 9, 29, 30, 32, 33, 34, 35, 36, 37, 38, 42, 60
water quality standards, 9, 32, 33, 36, 38
water supplies, 83, 140, 142, 147
watershed, 41
waterways, 32, 35, 112, 119
weapons, 8
websites, 103
welfare, 6, 35, 52, 60, 76, 91, 150
wetlands, 38
White House, 26
wildlife, 33, 140, 142, 147, 150
within 90 days, xiii, 136, 137, 138, 140, 141, 142
workplace, 54
worldwide, 125